Optical Thin Films and Structures

Optical Thin Films and Structures: Design and Advanced Applications

Editor

Tsvetanka Babeva

MDPI • Basel • Beijing • Wuhan • Barcelona • Belgrade • Manchester • Tokyo • Cluj • Tianjin

Editor
Tsvetanka Babeva
Bulgarian Academy of Sciences
Bulgaria

Editorial Office
MDPI
St. Alban-Anlage 66
4052 Basel, Switzerland

This is a reprint of articles from the Special Issue published online in the open access journal *Coatings* (ISSN 2079-6412) (available at: https://www.mdpi.com/journal/coatings/special_issues/optical_film_struct_des_appl).

For citation purposes, cite each article independently as indicated on the article page online and as indicated below:

LastName, A.A.; LastName, B.B.; LastName, C.C. Article Title. *Journal Name* **Year**, *Volume Number*, Page Range.

ISBN 978-3-0365-0892-4 (Hbk)
ISBN 978-3-0365-0893-1 (PDF)

© 2021 by the authors. Articles in this book are Open Access and distributed under the Creative Commons Attribution (CC BY) license, which allows users to download, copy and build upon published articles, as long as the author and publisher are properly credited, which ensures maximum dissemination and a wider impact of our publications.

The book as a whole is distributed by MDPI under the terms and conditions of the Creative Commons license CC BY-NC-ND.

Contents

About the Editor vii

Tsvetanka Babeva
Special Issue: "Optical Thin Films and Structures: Design and Advanced Applications"
Reprinted from: *Coatings* 2020, *10*, 1140, doi:10.3390/coatings10111140 1

Alexander Tikhonravov, Igor Kochikov, Ivan Matvienko, Temur Isaev and Anatoly Yagola
Strategies of Broadband Monitoring Aimed at Minimizing Deposition Errors
Reprinted from: *Coatings* 2019, *9*, 809, doi:10.3390/coatings9120809 5

Olaf Stenzel, Steffen Wilbrandt, Christian Mühlig and Sven Schröder
Linear and Nonlinear Absorption of Titanium Dioxide Films Produced by Plasma Ion-Assisted Electron Beam Evaporation: Modeling and Experiments
Reprinted from: *Coatings* 2020, *10*, 59, doi:10.3390/coatings10010059 15

Dimitre Dimitrov, Che-Liang Tsai, Stefan Petrov, Vera Marinova, Dimitrina Petrova, Blagovest Napoleonov, Blagoy Blagoev, Velichka Strijkova, Ken Yuh Hsu and Shiuan Huei Lin
Atomic Layer-Deposited Al-Doped ZnO Thin Films for Display Applications
Reprinted from: *Coatings* 2020, *10*, 539, doi:10.3390/coatings10060539 31

Akmedov K. Akhmedov, Aslan Kh. Abduev, Vladimir M. Kanevsky, Arsen E. Muslimov and Abil Sh. Asvarov
Low-Temperature Fabrication of High-Performance and Stable GZO/Ag/GZO Multilayer Structures for Transparent Electrode Applications
Reprinted from: *Coatings* 2020, *10*, 269, doi:10.3390/coatings10030269 43

Chi-Fan Liu, Chun-Hsien Kuo, Tao-Hsing Chen and Yu-Sheng Huang
Optoelectronic Properties of Ti-doped SnO_2 Thin Films Processed under Different Annealing Temperatures
Reprinted from: *Coatings* 2020, *10*, 394, doi:10.3390/coatings10040394 55

Dervil Cody, Tsvetanka Babeva, Violeta Madjarova, Anastasia Kharchenko, Sabad-e-Gul, Svetlana Mintova, Christopher J. Barrett and Izabela Naydenova
In-Situ Ellipsometric Study of the Optical Properties of LTL-Doped Thin Film Sensors for Copper(II) Ion Detection
Reprinted from: *Coatings* 2020, *10*, 423, doi:10.3390/coatings10040423 67

Katerina Lazarova, Silvia Bozhilova, Christo Novakov, Darinka Christova and Tsvetanka Babeva
Amphiphilic Poly(vinyl Alcohol) Copolymers Designed for Optical Sensor Applications—Synthesis and Properties
Reprinted from: *Coatings* 2020, *10*, 460, doi:10.3390/coatings10050460 79

Maarten Eerdekens, Ismael López-Duarte, Gunther Hennrich and Thierry Verbiest
Thin Films of Tolane Aggregates for Faraday Rotation: Materials and Measurement
Reprinted from: *Coatings* 2019, *9*, 669, doi:10.3390/coatings9100669 91

Xinfang Huang, Zhiwen Xie, Kangsen Li, Qiang Chen, Yongjun Chen and Feng Gong
Thermal Stability of CrWN Glass Molding Coatings after Vacuum Annealing
Reprinted from: *Coatings* 2020, *10*, 198, doi:10.3390/coatings10030198 97

Hongpeng Shang, Degui Sun, Peng Yu, Bin Wang, Ting Yu, Tiancheng Li and Huilin Jiang
Investigation for Sidewall Roughness Caused Optical Scattering Loss of Silicon-on-Insulator Waveguides with Confocal Laser Scanning Microscopy
Reprinted from: *Coatings* **2020**, *10*, 236, doi:10.3390/coatings10030236 . . . . **109**

About the Editor

Tsvetanka Babeva is currently a Full Professor and Director of the Institute of Optical Materials and Technology at the Bulgarian Academy of Sciences, Sofia, Bulgaria. She received her Ph.D. in Physics from the Bulgarian Academy of Science in 2003 and completed her habilitation in 2010. Her professional experience includes 24 years as a scientist at the Bulgarian Academy of Sciences and 2 years as an Arnold F. Graves fellow at the Centre of Industrial and Engineering Optics at Technical University Dublin, Ireland. Her present research is devoted to the design, preparation and optical characterization of thin films and structures and developing of porous materials and photonic structures for optical sensing applications.

Editorial

Special Issue: "Optical Thin Films and Structures: Design and Advanced Applications"

Tsvetanka Babeva

Institute of Optical Materials and Technologies "Akad. J. Malinowski", Bulgarian Academy of Sciences, Akad. G. Bonchev str., bl. 109, 1113 Sofia, Bulgaria; babeva@iomt.bas.bg; Tel.: +359-2-979-3502

Received: 11 November 2020; Accepted: 18 November 2020; Published: 23 November 2020

Abstract: This Special Issue is devoted on design and application of thin films and structures with special emphasis on optical applications. It comprises ten papers, five featured and five regular papers, authored by respective scientists all over the world. Diverse materials are studied and their possible applications are demonstrated and discussed: transparent conductive coatings and structures from ZnO doped with Al and Ga and Ti-doped SnO_2, polymer and nanosized zeolite thin films for optical sensing, TiO_2 with linear and non-linear optical properties, organic diamagnetic materials, broadband optical coatings, CrWN glass molding coatings and silicon on insulator waveguides.

Keywords: transparent conductive coatings; optical sensing; broadband design; linear and non-linear optical properties; sidewall roughness; organic diamagnetic materials

1. Introduction

Diverse types of materials such as polymers, glasses, metals, ceramics, zeolites, etc., could be prepared as thin films with high optical quality thus finding applications in photonics, optical sensing, photocatalysis, optoelectronics, linear and non-linear optics, holography, etc. Different production strategies, including "dry" and "wet" deposition methods, are developed and optimized. In order for these thin films and structures to be utilized in different optical devices, unambiguous methods for design and characterization are required. Additionally, in-situ optical monitoring of their properties will be beneficial for proper device operation.

This Special Issue covers the recent progress and new developments in the area of design, deposition, characterization and application of optical thin films and structures.

2. Statistics of the Special Issue

The special issue consists of 10 full papers authored by 57 authors. The geographical distribution of authors can be seen in Figure 1. The authors originate from 10 countries from three different continents—Europe, Asia and North America. The average number of authors per manuscript is 5.7.

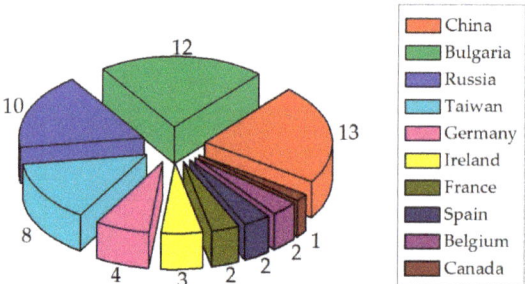

Figure 1. Geographic distribution by the country of authors.

3. Brief Overview of the Contributions to This Special Issue

Tikhonravov et al. [1] presented a computational approach for comparing various broadband monitoring strategies, taking into account the positive and negative effects associated with the correlation of thickness errors caused by the monitoring procedure. The presented computational approach is general and can be applied to check the prospects of the production of various types of optical coatings. Stenzel et al. [2] studied linear and non-linear optical properties of titanium dioxide films prepared by plasma ion-assisted electron beam evaporation. Linear optical properties were investigated in terms of spectrophotometry using the beta-distributed oscillator model as a parametrized dispersion law. The nonlinear two-photon absorption coefficient of titanium dioxide was determined by means of the laser-induced deflection technique at a wavelength of 800 nm. Dimitrov et al. [3] demonstrated transparent and conductive aluminum-doped zinc oxide (AZO) thin films deposited on rigid and transparent substrates through atomic layer deposition. Applications as transparent conductive layers in AZO/glass-supported liquid crystal displays and flexible polymer-dispersed liquid crystal devices were discussed. Akhmedov et al. [4] investigated the structural, electrical, and optical performances of Ga-doped ZnO/Ag/Ga-doped ZnO (GZO/Ag/GZO) multilayered structures deposited on glass substrates by direct current (DC) magnetron sputtering in a pure Argon medium without any substrate heating. Highly transparent and conductive samples were obtained. Liu et al. [5] investigated Ti-doped SnO_2 transparent conductive oxide thin films deposited on glass substrates using radio frequency (RF) magnetron sputtering and postdeposition annealing at temperatures in the range of 200–500 °C for 30 min. The effects of the annealing temperature on the structural properties, surface roughness, electrical properties, and optical transmittance of the thin films are then systematically explored. Cody et al. [6] demonstrated a possibility of optical sensing of copper ions in water using Linde Type L (LTL) zeolite thin films. Both single wavelength and spectroscopic ellipsometry were used for characterization of the changes in optical constants and thickness of films in the presence of heavy metal ions. Lazarova et al. [7] demonstrated a possible approach for enhancement of Poly(vinyl alcohol) (PVA) humidity-sensing ability using poly(vinylalcohol-*co*-vinylacetal) copolymers of different acetal content. Further enhancement through preparation of polymer–silica hybrids was demonstrated. The possibility of color sensing of humidity was also discussed. Eerdekens et al. [8] demonstrated organic, diamagnetic materials based on structurally simple (hetero-)tolane derivatives that form crystalline thin-film aggregates suitable for Faraday rotation spectroscopy. Huang at al. [9] studied the impact of vacuum annealing on CrWN glass molding coatings deposited by plasma enhanced magnetron sputtering. The vacuum annealing induced surface coarsening and spinodal decomposition accompanied by the formation of nm-sized c-CrN, c-W_2N, and h-WN domains. The large volume fraction of the last one seriously weakened the coating strength and caused a drop in hardness. Shang et al. [10] used a theoretical/experimental combinative model for investigation of the waveguide sidewall roughness (SWR) and its impact on the optical propagation losses in silicon-on-insulator waveguides.

4. Conclusions

In making this Special Issue on Optical Thin Films and Structures: Design and Advanced Applications, I had the pleasure of communicating with first-class authors worldwide and the chance to obtain high quality contributions. I am very grateful to all the authors of the Special Issue for their submissions. I hope that the papers will be useful and of interest for the readers.

Conflicts of Interest: The author declares no conflict of interest.

References

1. Tikhonravov, A.; Kochikov, I.; Matvienko, I.; Isaev, T.; Yagola, A. Strategies of broadband monitoring aimed at minimizing deposition errors. *Coatings* **2019**, *9*, 809. [CrossRef]
2. Stenzel, O.; Wilbrandt, S.; Mühlig, C.; Schröder, S. Linear and nonlinear absorption of titanium dioxide films produced by plasma ion-assisted electron beam evaporation: Modeling and experiments. *Coatings* **2020**, *10*, 59. [CrossRef]
3. Dimitrov, D.; Tsai, C.-L.; Petrov, S.; Marinova, V.; Petrova, D.; Napoleonov, B.; Blagoev, B.; Strijkova, V.; Hsu, K.Y.; Lin, S.H. Atomic Layer-Deposited Al-Doped ZnO Thin Films for Display Applications. *Coatings* **2020**, *10*, 539. [CrossRef]
4. Akhmedov, A.K.; Abduev, A.K.; Kanevsky, V.M.; Muslimov, A.E.; Asvarov, A.S. Low-temperature fabrication of high-performance and stable GZO/Ag/GZO multilayer structures for transparent electrode applications. *Coatings* **2020**, *10*, 269. [CrossRef]
5. Liu, C.-F.; Kuo, C.-H.; Chen, T.-H.; Huang, Y.-S. Optoelectronic properties of Ti-doped SnO_2 thin films processed under different annealing temperatures. *Coatings* **2020**, *10*, 394. [CrossRef]
6. Cody, D.; Babeva, T.; Madjarova, V.; Kharchenko, A.; Gul, S.; Mintova, S.; Barrett, C.J.; Naydenova, I. In-situ ellipsometric study of the optical properties of LTL-doped thin film sensors for Copper(II) ion detection. *Coatings* **2020**, *10*, 423. [CrossRef]
7. Lazarova, K.; Bozhilova, S.; Novakov, C.; Christova, D.; Babeva, T. Amphiphilic Poly(vinyl Alcohol) copolymers designed for optical sensor applications—synthesis and properties. *Coatings* **2020**, *10*, 460. [CrossRef]
8. Eerdekens, M.; López-Duarte, I.; Hennrich, G.; Verbiest, T. Thin films of tolane aggregates for Faraday rotation: Materials and measurement. *Coatings* **2019**, *9*, 669. [CrossRef]
9. Huang, X.; Xie, Z.; Li, K.; Chen, Q.; Chen, Y.; Gong, F. Thermal stability of CrWN glass molding coatings after vacuum annealing. *Coatings* **2020**, *10*, 198. [CrossRef]
10. Shang, H.; Sun, D.; Yu, P.; Wang, B.; Yu, T.; Li, T.; Jiang, H. Investigation for sidewall roughness caused optical scattering loss of silicon-on-insulator waveguides with confocal laser scanning microscopy. *Coatings* **2020**, *10*, 236. [CrossRef]

Publisher's Note: MDPI stays neutral with regard to jurisdictional claims in published maps and institutional affiliations.

© 2020 by the author. Licensee MDPI, Basel, Switzerland. This article is an open access article distributed under the terms and conditions of the Creative Commons Attribution (CC BY) license (http://creativecommons.org/licenses/by/4.0/).

Article

Strategies of Broadband Monitoring Aimed at Minimizing Deposition Errors

Alexander Tikhonravov [1],*, Igor Kochikov [1], Ivan Matvienko [1], Temur Isaev [2] and Anatoly Yagola [2]

1. Research Computing Center, Lomonosov Moscow State University, Leninskiye Gory, Moscow 119992, Russia; igor@kochikov.ru (I.K.); matvienko.ivan.a@gmail.com (I.M.)
2. Lomonosov Moscow State University, Faculty of Physics, Leninskiye Gory, Moscow 119991, Russia; temurisaev@gmail.com (T.I.); yagola@phys.msu.ru (A.Y.)
* Correspondence: tikh@srcc.msu.ru; Tel.: +7-916-684-7612

Received: 7 November 2019; Accepted: 27 November 2019; Published: 1 December 2019

Abstract: This article presents a computational approach for comparing various broadband monitoring strategies, taking into account the positive and negative effects associated with the correlation of thickness errors caused by the monitoring procedure. The approach is based on statistical estimates of the strength of the error self-compensation effect and the expected level of thickness errors. Its application is demonstrated by using a 50-layer, nonpolarizing edge filter. The presented approach is general and can be applied to verify the prospects of broadband monitoring for the production of various types of optical coatings.

Keywords: thin films; optical coatings; monitoring; deposition

1. Introduction

Optical monitoring techniques can be applied in the production of optical coatings in almost all deposition plants. Both commercial and homemade optical monitoring devices are widely used all over the world, and choosing a good optical monitoring strategy is a key issue for the production of high-quality optical coatings. There is a great variety of optical monitoring techniques, and they are divided into monochromatic and broadband techniques [1]. In the case of monochromatic techniques, the question of having a proper monitoring strategy was raised many decades ago. The most impressive example of this was the production of narrow band-pass filters using turning point optical monitoring [2–4]. The production of very complicated optical filters became possible due to the presence of a very strong error self-compensation effect associated with this type of monitoring. The physics of the error self-compensation effect was explained many years later [5], and it was shown that the advantage of turning point monitoring appears only in the case of filters with resonant filter cavities. For other types of optical coatings, monochromatic-level monitoring was proposed many years ago in various forms [6–9]. For monochromatic-level monitoring, the choice of monitoring strategy is usually decided by the specifications of monitoring wavelengths and signal termination levels or swing [10] termination levels for all coating layers. The choice of monochromatic monitoring strategy is not a straightforward task, and recent efforts have been made to automate this choice [11,12].

In the case of broadband monitoring, various monitoring strategies are also possible. The first alternative is the choice between direct and indirect monitoring strategies [1,10]. The main advantage of direct monitoring was indicated by Macleod [7]. In this case, we monitor one of the samples that we want to produce. Unfortunately, direct broadband monitoring can lead to the development of a strong, cumulative effect of thickness error growth. This effect was even noticed in the first works done on broadband monitoring [13,14] and was later investigated in detail [15]. Indirect broadband monitoring allows one to use several monitoring chips and, thus, prevent the fast growth of thickness errors.

However, with this type of monitoring, we lose the previously noted advantage of direct monitoring. The recent progress in monitoring hardware arrangements [16,17] allows one to combine the advantages of direct and indirect monitoring. In the arrangement reported in these works, the monitor holder has several monitoring chips and is located on the main wheel of the deposition chamber with the same radial position as those of the deposited samples. Thus, the cumulative effect of thickness errors can be reduced by using several monitoring chips during the coating deposition.

The negative cumulative effect of thickness error growth is connected with the correlation of thickness errors by optical monitoring procedures [15]. Although, the correlation of errors can also lead to a positive effect of self-compensation of influence of errors in various layer thicknesses. In the case of broadband monitoring, this effect was first noticed four decades ago [13,14]. However, a comprehensive study of the error self-compensation effect associated with direct broadband monitoring began only recently, after the presence of a very strong effect was detected in the production of Brewster's angle polarizers for high-fluence optics [18]. The mathematical investigation of the error self-compensation mechanism in the case of broadband monitoring was performed in Ref. [19]. The results of this investigation were formulated in terms of singular values of rectangular matrices describing the correlation of errors in the course of broadband monitoring. Unfortunately, this form of representation is not convenient for practical applications, and the degree of correlation between the thickness errors and the strength of the error self-compensation effect can be calculated using computational experiments on optical coating production simulations [20,21].

The recent progress in broadband monitoring hardware [16,17] allows one to apply different direct monitoring strategies. In particular, direct broadband monitoring can be performed using several monitoring chips. It is also possible to remove monitoring chips and bring them back to the measurement position many times during the coating deposition. This strategy was previously applied in the case of indirect broadband monitoring, and it was shown that it had a certain advantage in monitoring some types of optical coatings [22].

Despite the obvious progress in monitoring arrangement, choosing the optimal strategy in the case of broadband monitoring is still an open question. When studying this question, we should take into account the negative and positive effects connected with the correlation of thickness errors. On the one hand, the use of several monitoring chips prevents rapid development of the cumulative effect of thickness errors. On the other hand, it can also reduce the degree of correlation of thickness errors and the associated positive effect of error self-compensation. The goal of this paper is to present a computational approach that can be applied for the comparison of various monitoring strategies while also taking into account the above-mentioned effects. We hope that such a comparison will be useful in practice to help select the optimal monitoring strategy for a given coating design.

2. The Computational Approach to Assessing the Degree of Thickness Error Correlation and the Strength of the Error Self-Compensation Effect

To illustrate the proposed computational approach, we analyzed a design of 50-layer, nonpolarizing edge filter with a 45° light incidence. Its theoretical spectral characteristics and layer physical thicknesses are presented in Figure 1. The filter used model high- and low- index materials with refractive indices of 2.35 and 1.45 (for example, model TiO_2 and SiO_2 indices). The first layer counting from the substrate was the high-index material layer, and the substrate refractive index was 1.52. It was designed using OptiLayer thin film software (v12.12) [23]. Computational manufacturing experiments with this filter [24] demonstrated the presence of a strong error self-compensation effect in the case of broadband monitoring in the normal incidence transmittance mode.

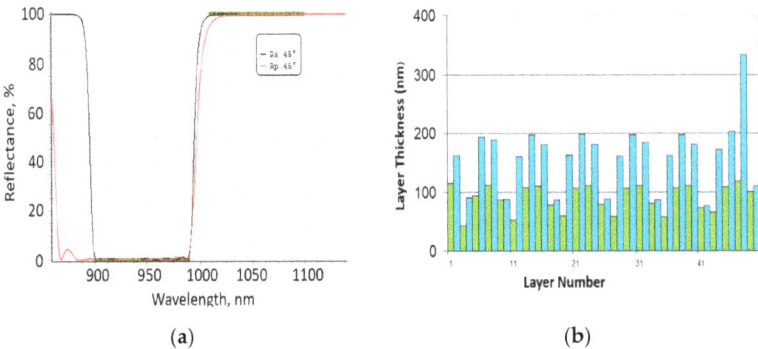

Figure 1. Theoretical *s*- and *p*-reflectances (**a**) and thicknesses (**b**) of the 50-layer, nonpolarizing edge filter.

Let $d_1^t, \ldots, d_m^t$ be physical thicknesses of a coating design. Here, m is the total number of coating layers, and $m = 50$ in the case of the considered nonpolarizing filter. In the course of production, actual layer thicknesses $d_1^a, \ldots, d_m^a$ differed from the planned values. Consider broadband monitoring using measured transmittance spectra. When using modern broadband monitoring devices, such spectra usually have hundreds or even thousands of spectral points. Let d be the growing thickness of the j-th coating layer. The measured transmittance is

$$T_j(d) = T_j(d_1^a, \ldots, d_{j-1}^a, d) + \delta T_{\text{meas}} \tag{1}$$

Here, δT_{meas} is the error in measured transmittance data. In Equation (1) and the following equations, we omitted the indication of an obvious dependence on wavelength λ.

With broadband monitoring, the deposition of the j-th layer is terminated in accordance with the condition that the minimum is reached by the discrepancy function

$$\Phi_j(d) = \sum_\lambda \left[T_j(d_1^a, \ldots, d_{j-1}^a, d) + \delta T_{\text{meas}} - T_j(d_1^t, \ldots, d_j^t) \right]^2 \to \min \tag{2}$$

Here, the summation is carried out over the wavelength grid at which the transmittance is measured.

It follows from Equation (2) that the actual thickness of the deposited j-th layer is associated not only with the errors in transmittance data but is also determined by the actual thicknesses of all previously deposited layers. This is the reason for the correlation of errors in layer thicknesses.

As outlined above, a rigorous mathematical investigation of the correlation of thickness errors was provided in [19]. To present the main result of this investigation, we introduced the vector of thickness errors $\Delta = \{\delta d_1, \ldots, \delta d_m\}$. When considering Equation (2) for all coating layers, starting from layer $j = 2$, the following matrix appears:

$$S_j = \left\| \sum_\lambda \frac{\partial T_j}{\partial d_i} \frac{\partial T_j}{\partial d_k} \right\|. \tag{3}$$

Here, $\partial T_j / \partial d_i$ are partial derivatives of the intensity transmission coefficient for the subsystem of layers with the numbers from 1 to j.

Let λ_i^j and P_j^i be eigenvalues and eigenvectors of the matrix Cj, and $p_1^{ij}, ..., p_j^{ij}$ be the elements of the eigenvectors P_j^i. With their help, the following raw vectors are introduced for all i from 1 to j and all j from 2 to m:

$$W_{ij} = \sqrt{\lambda_i^j} \{p_1^{ij}, ..., p_j^{ij}, 0, .., 0\}. \tag{4}$$

These raw vectors were then used to form the rectangular matrix W with the dimensions $k \times m$, where m is the number of coating layers, and $k = (m − 1)/(m + 2)/2$. In accordance with the results of Ref. [19], the correlation of thickness errors led to a small norm of the vector WΔ. To formalize the concept of the smallness of this norm, the parameter α was introduced in [20]. It is calculated as

$$\alpha = \|W\Delta^0\|^2 / \langle \|W\Delta^r\|^2 \rangle \tag{5}$$

Here, Δ^0 is the normalized error vector Δ (i.e., $\Delta^0 = \Delta / \|\Delta\|$), and $\langle \|W\Delta^r\|^2 \rangle$ is the averaged square norm of the vectors $W\Delta^r$ over all vectors Δ^r with the norm $\|\Delta^r\| = 1$.

The introduced parameter α compares the value of the norm $W\Delta^0$ with the average value of the norm $W\Delta^r$ for all random vectors of the unit length. The introduced parameter is called the degree of thickness error correlation. The smaller this parameter the stronger the correlation of errors.

In [20,21], the degree of thickness error correlation was estimated for two cases in which a strong error self-compensation effect was observed either practically [18] or in the course of simulating the coating deposition [24]. The Brewster's angle polarizer and the nonpolarizing edge filter were considered here. In both cases, the smallness of parameter α was confirmed. We consider the related results for the edge filter later in this document.

Subsequently, we proceeded to assess the strength of the error self-compensation effect. Despite the correlation of thickness errors, error vectors Δ are also random in nature since they are determined by various random factors. Therefore, the estimation of the strength of the error self-compensation effect should have a statistical form. Since this effect is caused by the correlation of thickness errors, it is also natural to estimate it by comparing it with the influence of uncorrelated thickness errors.

To assess the impact of errors on coating spectral characteristics, we used the merit function to solve the respective design problem. In the case of the nonpolarizing edge filter, it has the form

$$MF = \left\{ \frac{1}{2L} \sum_\lambda \left[(T_s - \hat{T})^2 + (T_p - \hat{T})^2 \right] \right\}^{1/2} \tag{6}$$

Here, T_s and T_p are transmittances for the s- and p-polarized light at 45° light incidence; $\hat{T}$ is the target transmittance, equal to 0% in the 900–990 nm spectral band and equal to 100% in the 1010–1100 nm band; and L is the total number of spectral grid points that has the step of 1 nm in the two target spectral bands.

Let us designate $\delta MF(\Delta)$ as the deviation of merit function corresponding to the error vector Δ. We wanted to compare the deviations that are caused by the correlated and uncorrelated thickness errors.

The uncorrelated thickness errors are most consistent with stable production processes using time and quartz crystal monitoring. It is generally accepted that when using these monitoring techniques, the best accuracy in controlling layer thicknesses is about 1% of the planned thickness values. Let us denote $\langle \delta MF \rangle$ as the root-mean-square value of merit function deviations calculated for the large number of random error vectors Δ that are set so their coordinates Δd_j are distributed according to Gaussian law with zero mathematical expectations and standard deviations equal to 1% of the thicknesses of the corresponding layers. Thus, $\langle \delta MF \rangle$ is an estimate of the effect of uncorrelated errors. Further, to obtain this estimate, we generated 1 million uncorrelated error vectors.

To obtain the vector of correlated thickness errors, one can use computational manufacturing experiments to simulate the deposition process with broadband optical monitoring. In [24], OptiLayer software [23] was used for this purpose. In [21], a simplified simulator of this process was proposed that, on the one hand, fully reflected the process of thickness error correlation and, on the other hand, allowed error vectors to be generated much faster than full simulators of the deposition processes. In this paper, we applied the simplified simulator from [21].

In order to evaluate the strength of the error self-compensation effect S, two possible assessments were considered in our previous works [20,25]. They were also based on a comparison of the effect of correlated and uncorrelated thickness errors. In Ref. [20], a special D_α region was considered in the unit sphere in the space of error vectors, and an estimate for parameter S was introduced using all correlated error vectors with a degree of correlation of thickness errors less than α. In Ref. [25], an estimate for S was introduced with the normalization of all error vectors to unit norm vectors. Here, we introduce a new evaluation form for S, which we hope is more consistent with practice.

Depending on the levels of simulated error factors, generated error vectors Δ will have various norms. In general, with a lower norm of the error vector, lower values of the merit function variations should be expected. For a more objective comparison of correlated and uncorrelated thickness errors, it is advisable to consider, in both cases, error vectors of the same norm. For this reason, we normalized all correlated errors vectors so that their norms were 1% of the design vector norm. With this normalization, the strength of the error self-compensation effect for a specific vector of correlated errors will be estimated as

$$S = \langle \delta MF \rangle / \delta MF(\Delta) \tag{7}$$

Figure 2 shows the probability density functions of the distributions of the degree of correlation of thickness errors and the strength of the error self-compensation effect in the case of direct transmittance monitoring in the 450–950 nm spectral band. These distributions are calculated using 1,000,000 vectors of correlated errors.

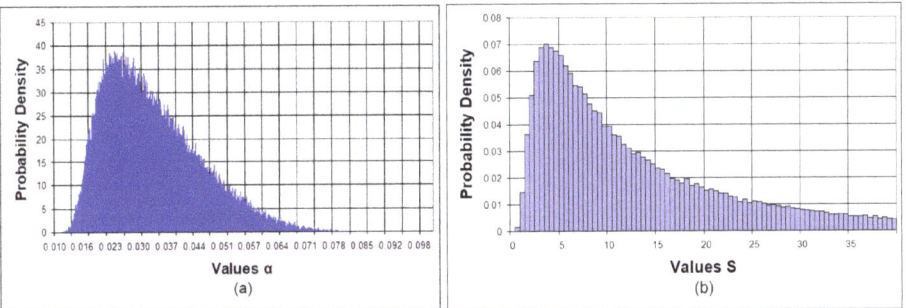

Figure 2. Probability density functions for the degree of thickness error correlation α (**a**) and for the strength of the error self-compensation effect S (**b**).

In full accordance with previously obtained results [18,24], Figure 2 demonstrates the smallness of parameter α. At the same time, almost all calculated S values were large enough, which indicates the presence of a strong error self-compensation effect. The average S value was equal to 16.6.

An even more visual representation of a strong correlation of thickness errors and the associated error self-compensation effect is given by Figure 3, where pairs of α and S values are presented for correlated and uncorrelated error vectors. The pairs of values corresponding to these two types of errors are located in significantly different parts of the (α, S) plane.

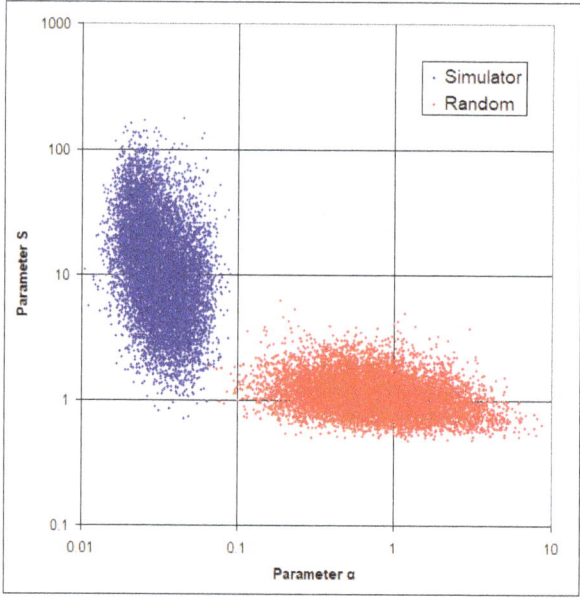

Figure 3. Comparison of pairs α and S for correlated (blue) and uncorrelated (red) error vectors. Dots represent 10,000 tests randomly selected from calculated sequences of 1,000,000 vectors.

3. Comparison of Various Monitoring Strategies

In this section, we used the estimates of the previous section to compare four different monitoring strategies. The first one was the direct monitoring strategy with all filter layers monitored using one of the samples to be produced. The next strategy used two subsequent monitoring chips, so that filter layers with the numbers from 1–24 were monitored using the first chip, and layers 25–50 were monitored using the second chip. The third strategy applied four monitoring chips that were used to monitor layers 1–12, 13–24, 25–36, and 37–50. The fourth strategy used two chips that were moved out of the measurement position and returned back to this position so that the first chip monitored layers 1, 2, 4, ... , 50, while the second chip monitored layers 3, 5, ... , 49. Monitoring the first two layers with a single chip allowed us to increase the optical contrast for monitoring low-index layers (even layers) by applying the first high-index layer to this chip. All four strategies caused correlation of thickness errors, but we expected that, in the first case, parameter α would have lower values than in the other cases. Recall that parameter α is smaller when the correlation of errors is higher.

In all four cases, 1,000,000 error vectors were generated to calculate α and S values. Figure 4 shows the probability density functions for the degree of thickness error correlation α and the strength of the error self-compensation effect S.

As one may expect, parameter α had lower values in the case of direct monitoring. This reflects a stronger correlation of thickness errors when all layer thicknesses are monitored using a single sample. In the case of direct monitoring, the average S value was the largest, and it was equal to 16.6.

For predictive comparisons of various monitoring strategies, we also needed to compare thickness error levels for all four cases. Recall that for the comparison with uncorrelated errors, all error vectors in Equation (7) were normalized to the same value. Following this, strategies with several monitoring chips were introduced to reduce thickness error levels. Figure 5 shows the probability density functions of the distributions of norms of error vectors for the considered strategies. As before, calculations were performed based on 1,000,000 simulation tests in each case.

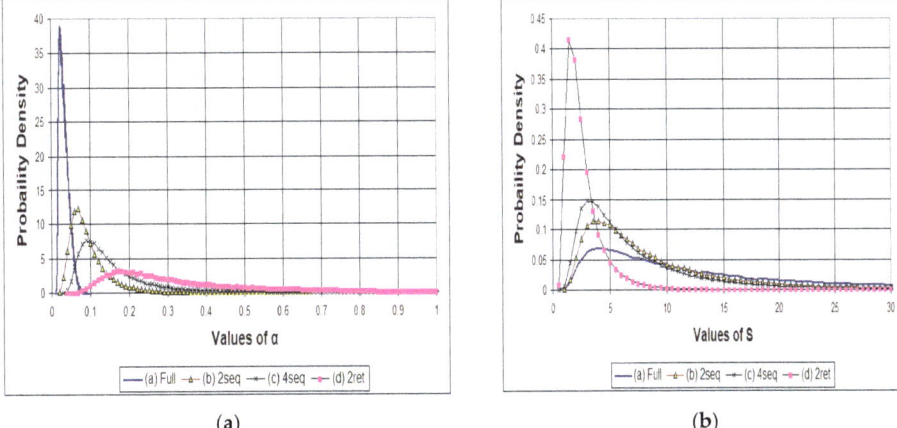

Figure 4. Probability density functions for α (**a**) and *S* (**b**) in the cases of different monitoring strategies. Full—direct strategy, 2seq—strategy with two subsequent chips, 4seq—strategy with four subsequent chips, and 2ret—strategy with two returning chips.

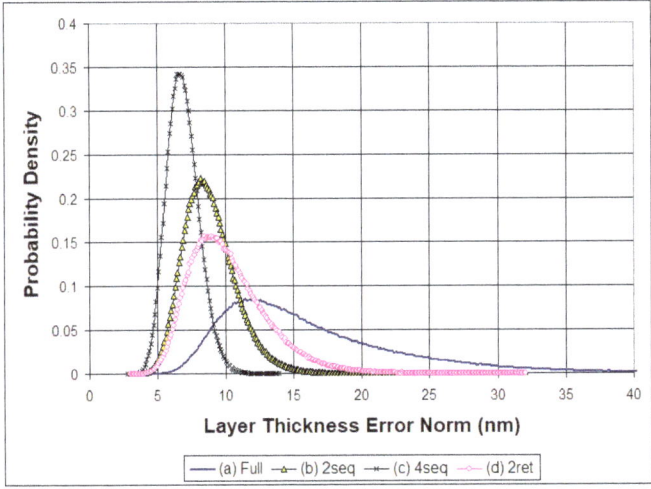

Figure 5. Probability density functions for the norms of error vectors: (**a**) direct strategy, (**b**) strategy with two subsequent chips, (**c**) strategy with four subsequent chips, and (**d**) strategy with two returning chips.

Indeed, levels of thickness errors were noticeably reduced when the strategies with several monitoring chips were applied. The average values of error vector norms in Figure 5 were equal to 17.25 nm in the case of direct monitoring and 9.32, 6.92, and 10.55 nm in the cases of strategies with several monitoring chips.

Despite somewhat weaker error self-compensation effects, the strategies with several monitoring chips may be preferable because of the lower levels of thickness errors. To evaluate a positive effect of error self-compensation, taking into account the expected levels of thickness errors, we made the following considerations. From a theoretical point of view, in the first approximation, the deviation $\delta MF(\Delta)$ grew linearly with an increase in the norm of the error vector Δ. The strength of the error self-compensation effect S was estimated by Equation (7) for the error vectors Δ with the norms equal to $0.01D$, where D is the norm of the design vector. Let us denote as $\langle \|\Delta\| \rangle$ the average values of the error vector norms in the distributions shown in Figure 5. Using these average values, we introduced the effective strength of the error self-compensation effect for a given monitoring strategy by the equation

$$S_{eff} = \langle S \rangle \frac{0.01D}{\langle \|\Delta\| \rangle} \qquad (8)$$

Table 1 compares average S values in Figure 4b, average $\|\Delta\|$ in Figure 5, and S_{eff} values for the considered four monitoring strategies.

Table 1. Average S, average norm $\|\Delta\|$, and S_{eff} values for the four discussed monitoring strategies.

Strategy	$\langle S \rangle$	$\langle \|\Delta\| \rangle$	S_{eff}
direct	16.55	17.25	9.32
2 subs. chips	9.42	9.32	9.82
4 subs. chips	7.38	6.92	10.36
2 returning chips	2.48	10.55	2.28

The discussion of the obtained results is provided in the next section.

4. Discussion

Recent achievements in the development of broadband monitoring hardware allow one to combine advantages of direct and indirect monitoring strategies through the use of several monitoring chips that are located on the main wheel of the deposition chamber. Even more, it is also possible to remove monitoring chips and bring them back to the measurement position many times during the coating deposition. This opens a way for using various broadband monitoring strategies. Thus, the question of comparing various strategies and choosing the most appropriate one becomes important. The presented research outlines a way for answering this question.

When considering optical monitoring strategies, we should take into account the correlation of thickness errors by optical monitoring procedures. This correlation causes both negative and positive effects. On one hand, it can lead to the development of a strong cumulative effect of thickness error growth, but on the other hand, it can result in the self-compensation of thickness errors. In this paper, we proposed a computational approach to assess the degree of thickness error correlation and the strength of the error self-compensation effect. The proposed approach was used to compare four strategies of broadband monitoring. It was shown that in the case of a 50-layer, nonpolarizing edge filter, the direct monitoring strategy provided the strongest correlation of thickness errors and the strongest error self-compensation effect. At the same time, in this case, one should expect the highest level of thickness errors caused by the negative cumulative effect of error growth. This reduces the effective strength of the error self-compensation effect. In the case of monitoring strategies with two and four subsequent monitoring chips, the strength of the error self-compensation effect was lower, but the expected levels of thickness errors were also lower. To evaluate the combined effect caused by the correlation of thickness errors, the effective strength of the error self-compensation effect S_{eff} was introduced by Equation (8). Table 1 shows that, in the case of strategies with several subsequent monitoring chips, S_{eff} was a bit higher than in the case of direct monitoring. However, taking into account the approximate nature of statistical estimates, on this basis, one should not conclude that the strategies with several subsequent chips have an absolute advantage in the case of a nonpolarizing edge filter. In the case of this design, all of the first three strategies deserve attention.

As for the fourth considered strategy, in the case of the nonpolarizing edge filter, it was clearly less suitable than the first three. However, this does not mean that the fourth strategy cannot be the best option for other designs. It is worth noting that the advantage of this strategy was discovered earlier [22] for a design with layers that were essentially thinner than the layers of the discussed filter.

The presented computational approach to comparing various broadband monitoring strategies is general and can be applied to check the prospects of the production of various types of optical coatings.

Author Contributions: Conceptualization, A.T.; Methodology, A.T.; Software, T.I. and I.M.; Validation, I.K. and I.M.; Formal Analysis: A.T., I.K., and A.Y.; Investigation: A.T. and T.I.; Writing—Original Draft Preparation, A.T.; Writing—Review and Editing, A.T. and I.K.; Funding Acquisition, A.T. and A.Y.; Supervision, A.T. and A.Y.

Funding: This research was funded by a grant from the Russian Scientific Foundation (No. 16-11-10219).

Conflicts of Interest: The authors declare no conflicts of interest.

References

1. Tikhonravov, A.; Trubetskov, M.; Amotchkina, T. Optical monitoring strategies for optical coating manufacturing. In *Optical Thin Films and Coatings*, 2nd Ed.; Piegari, A., Flory, F., Eds.; Woodhead: Cambridge, UK, 2018; pp. 65–101.
2. Bousquet, P.; Fornier, A.; Kowalczyk, R.; Pelletier, E.; Roche, P. Optical filters: Monitoring process allowing the autocorrection of thickness errors. *Thin Solid Films* **1972**, *13*, 285–290. [CrossRef]
3. Macleod, H. Turning value monitoring of narrow band optical filters. *Optica Acta Int. J. Opt.* **1972**, *19*, 1–28. [CrossRef]
4. Macleod, H.; Pelletier, E. Error compensation mechanisms in some thin-film monitoring systems. *Optica Acta Int. J. Opt.* **1977**, *24*, 907–930. [CrossRef]
5. Tikhonravov, A.V.; Trubetskov, M.K. Automated design and sensitivity analysis of wavelength-division multiplexing filters. *Appl. Opt.* **2002**, *41*, 3176–3182. [CrossRef]
6. Holm, C. Optical thin film production with continuous reoptimization of layer thicknesses. *Appl. Opt.* **1978**, *18*, 1978–1982. [CrossRef]
7. Macleod, H.A. Monitoring of optical coatings. *Appl. Opt.* **1981**, *20*, 82–89. [CrossRef]
8. Zhao, F. Monitoring of periodic multilayers by the level method. *Appl. Opt.* **1985**, *24*, 3339–3342. [CrossRef]
9. Laan, C. Optical monitoring of nonquarterwave stacks. *Appl. Opt.* **1986**, *25*, 757–760.
10. Macleod, H.A. *Thin-Film Optical Filters*, 4th ed.; CRC Press: Boca Raton, FL, USA; Taylor & Francis Group: Abingdon, UK, 2010.
11. Trubetskov, M.; Amotchkina, T.; Tikhonravov, A. Automated construction of monochromatic monitoring strategies. *Appl. Opt.* **2015**, *54*, 1900–1909. [CrossRef]
12. Vignaux, M.; Lemarchand, F.; Begou, T.; Grezes-Besset, C.; Lumeau, J. Semi-automated method for the determination of the all-optical monitoring strategy of complex thin-film filters. *Opt. Express* **2019**, *27*, 12373–12390. [CrossRef]
13. Vidal, B.; Fornier, A.; Pelletier, E. Optical monitoring of nonquarterwave multilayer filters. *Appl. Opt.* **1978**, *17*, 1038–1047. [CrossRef] [PubMed]
14. Vidal, B.; Fornier, A.; Pelletier, E. Wideband optical monitoring of nonquarterwave multilayer filters. *Appl. Opt.* **1979**, *18*, 3851–3856. [CrossRef] [PubMed]
15. Tikhonravov, A.; Trubetskov, M.; Amotchkina, T. Investigation of the effect of accumulation of thickness errors in optical coating production using broadband optical monitoring. *Appl. Opt.* **2006**, *45*, 7026–7034. [CrossRef] [PubMed]
16. Zoeller, A.; Williams, J.; Hartlaub, S. Precision filter manufacture using direct optical monitoring. In Proceedings of the OIC 11th Topical Meeting, OSA, Washington, DC, USA, 6–11 June 2010. TuC8.
17. Zoeller, A.; Hagedorn, H.; Weinrich, W.; Wirth, E. Testglass changer for direct optical monitoring. *Proc. SPIE* **2011**, *8168*, 81681J.
18. Zhupanov, V.; Kozlov, I.; Fedoseev, V.; Konotopov, P.; Trubetskov, M.; Tikhonravov, A. Production of Brewster-angle thin film polarizers using ZrO_2/SiO_2 pair of materials. *Appl. Opt.* **2017**, *56*, C30–C34. [CrossRef]
19. Tikhonravov, A.V.; Kochikov, I.V.; Yagola, A.G. Mathematical investigation of the error self-compensation mechanism in optical coating technology. *Inverse Probl. Sci. Eng.* **2018**, *26*, 1214–1229. [CrossRef]
20. Tikhonravov, A.V.; Kochikov, I.V.; Matvienko, I.A.; Sharapova, S.A.; Yagola, A.G. Estimates related to the error self-compensation mechanism in optical coatings deposition. *Mosc. Univ. Phys. Bull.* **2018**, *73*, 627–631. [CrossRef]
21. Tikhonravov, A.V.; Kochikov, I.V.; Matvienko, I.A.; Isaev, T.F.; Lukyanenko, D.V.; Sharapova, S.A.; Yagola, A.G. Correlation of errors in optical coating production with broad band monitoring. *Num. Methods Program.* **2018**, *19*, 439–448.

22. Zhupanov, V.; Klyuev, E.; Alekseev, S.; Kozlov, I.; Trubetskov, M.; Kokarev, M.; Tikhonravov, A. Indirect broadband optical monitoring with multiple witness substrates. *Appl. Opt.* **2009**, *48*, 2315–2320. [CrossRef]
23. OptiLayer. Thin Film Software. Available online: https://www.optilayer.com (accessed on 28 November 2019).
24. Tikhonravov, A.V.; Kochikov, I.V.; Yagola, A.G. Investigation of the error self-compensation effect associated with direct broad band monitoring of coating production. *Opt. Express* **2018**, *26*, 24964–24972. [CrossRef]
25. Tikhonravov, A.V.; Kochikov, I.V.; Sharapova, S.A. Broad band optical monitoring in the production of gain flattening filters for telecommunication applications. *Mosc. Univ. Phys. Bull.* **2019**, *74*, 160–164. [CrossRef]

© 2019 by the authors. Licensee MDPI, Basel, Switzerland. This article is an open access article distributed under the terms and conditions of the Creative Commons Attribution (CC BY) license (http://creativecommons.org/licenses/by/4.0/).

Article

Linear and Nonlinear Absorption of Titanium Dioxide Films Produced by Plasma Ion-Assisted Electron Beam Evaporation: Modeling and Experiments

Olaf Stenzel [1,*], Steffen Wilbrandt [1], Christian Mühlig [2] and Sven Schröder [1]

[1] Fraunhofer Institute of Applied Optics and Precision Engineering, IOF, Albert-Einstein-Str. 7, 07745 Jena, Germany; steffen.wilbrandt@iof.fraunhofer.de (S.W.); sven.schroeder@iof.fraunhofer.de (S.S.)
[2] Leibniz Institute of Photonic Technology, Albert-Einstein-Str. 9, 07745 Jena, Germany; christian.muehlig@leibniz-ipht.de
* Correspondence: olaf.stenzel@iof.fraunhofer.de; Tel.: +49-3641-807-348

Received: 20 December 2019; Accepted: 7 January 2020; Published: 9 January 2020

Abstract: Titanium dioxide films were prepared by plasma ion-assisted electron beam evaporation. Linear optical properties were investigated in terms of spectrophotometry using the beta-distributed oscillator (ß_do) model as a parametrized dispersion law. The nonlinear two-photon absorption coefficient of titanium dioxide was determined by means of the laser-induced deflection technique at a wavelength of 800 nm. The obtained values of $(2–5) \times 10^{-11}$ cm/W were consistent with published experimental values for rutile as well as for simulations performed in the frames of the ß_do and Sheik–Bahae models.

Keywords: optical coatings; titanium dioxide; optical constants; two-photon absorption; nonlinear refraction; scattering; laser-induced deflection; absorption measurement

1. Introduction

The standard theoretical apparatus used for modeling the optical properties of multilayer systems is formulated in terms of linear optics, i.e., it is based on a linear relationship between the electric field strength and the polarization induced in the medium. Within this framework, commonly used calculation recipes, such as matrix formalism or the admittance approach, have been derived in relevant textbooks [1–4]. Practical applications for these approaches in optical coatings design, characterization, and reengineering tasks have formed the content of relevant monographs (see, for example, References [2–5]).

Nevertheless, in high-power laser systems, the electric field in the incident light beam may be strong enough to induce relevant nonlinear optical effects in the coating materials. In many cases, it is the third-order (cubic) optical nonlinearity that dominates the nonlinear response. The optical Kerr effect, as well as nonlinear two-photon absorption (TPA), are prominent examples of third-order processes [6].

In high-power laser applications, third-order nonlinearity may therefore have to be taken into account in order to correctly predict the optical properties of a coating [7]. Thus, Razskazovskaya et al. [8] demonstrated the effect of TPA on the reflectance of dielectric laser mirrors in a pre-damage regime. Similarly, the optical Kerr effect appeared to be responsible for an intensity-dependent shift of the rejection band edge in several Angstroms in edge filters [9]. As opposed to linear optics, the general effect is that coating reflectance (and transmittance) becomes intensity-dependent. As we have shown in a previous study, these effects can principally be incorporated into a manageable design algorithm [10] when describing each coating material by four (instead of two) optical constants, namely linear and nonlinear refractive indexes as well as linear and nonlinear extinction or absorption coefficients. Concerning linear optical constants, an overwhelming number of studies exist that give an idea on the range of optical constants achievable for practically any relevant coating material depending on the deposition technique and parameters used. However, although there exist several manageable theoretical approaches for estimating nonlinear optical constants (see, for example, References [11–16]), reliable experimental data on the dispersion of nonlinear optical constants of thin-film materials are practically not available.

The motivation of this study was to contribute to an improvement of the data basis for nonlinear optical constants. We present results from the measurement of the two-photon absorption coefficient of TiO_2 thin optical films at a wavelength of 800 nm and discuss the results with respect to the predictions of two different theoretical approaches.

2. Theoretical Background

Both nonlinear refraction and absorption may be (in terms of the lowest order with respect to the field strength) expressed in terms of the cubic nonlinear dielectric susceptibility $\chi^{(3)}$. General quantum-mechanical expressions for $\chi^{(3)}$ are available in relevant textbooks (compare Reference [6]), but because of the high number of input parameters, their direct usage may be inconvenient in practical applications. Instead, we will make use of simplified expressions that allow for estimating the TPA coefficient β in terms of input parameters that are available from standard (linear) optical thin-film characterization techniques, among them spectrophotometry and ellipsometry.

In this context, the model of Sheik–Bahae et al. [15] provides a simple formula for calculating TPA contributions to nonlinear absorption in semiconducting solids. The material-specific input parameters of the model are the band gap E_g as well as the usual linear refractive index n. Both these values may be obtained from linear optical spectroscopy. When choosing the remaining, rather material-independent model parameters corresponding to what is recommended in Reference [15], the expression for β may be written as

$$\beta(\nu) \approx \frac{3100\sqrt{21}}{2^5 n^2 E_g^3} \frac{\left(\frac{2hc\nu}{E_g} - 1\right)^{\frac{3}{2}}}{\left(\frac{2hc\nu}{E_g}\right)^5}. \tag{1}$$

Equation (1) yields β directly in cm/GW, provided that both $hc\nu$ and E_g are given in eV (in which h represents Planck's constant, c is the velocity of light in a vacuum, and ν is the wavenumber, i.e., the reciprocal value of the light wavelength in a vacuum. In order to better understand the parametrization, Figure 1 shows estimated β values corresponding to Equation (1) for the crystalline TiO_2 modifications of anatase, rutile, and brookite. The underlying parameters are summarized in Table 1. The crystal refractive indices indicated in Table 1 are polarization-averaged.

Table 1. Mass density, band gap energy, and refractive index of selected TiO_2 modifications.

Modification	Mass Density gcm^{-3}	Band Gap E_g eV	E_g/hc cm^{-1}	$E_g/2hc$ cm^{-1}	Refractive Index n@589 nm
anatase	3.79–3.97	3.23	26,050	13,025	2.537
brookite	4.08–4.18	3.14	25,324	12,662	2.623
rutile	4.23	3.02	24,353	12,177	2.709
PIAD TiO_2 film	≈3.70	3.33	26,850	13,425	2.33
IBS TiO_2 film [17,18]	3.93	3.26	26,291	13,146	2.45

Note that, according to this model calculation, the main difference between the different modifications is in the onset energy of the TPA processes. The maximum β value is residually influenced by the choice of the modification, because material modifications with a lower gap tend to have a higher refractive index (compare Figure 2, where this trend is represented graphically for the sake of clarity), such that in Equation (1), the changes in n and E_g tend to cancel each other out. For reference purposes, we also include data from two thin-film samples, namely a sample produced by plasma ion-assisted evaporation (PIAD) as well as an ion beam-sputtered (IBS) sample, the latter prepared at Laser Zentrum Hannover LZH (for details, see Reference [17]). IBS is known as a preparation method that yields high-quality optical films with a rather high mass density. Indeed, the IBS data (Figure 2) fall closer to the values reported for the crystalline modifications than the PIAD sample does.

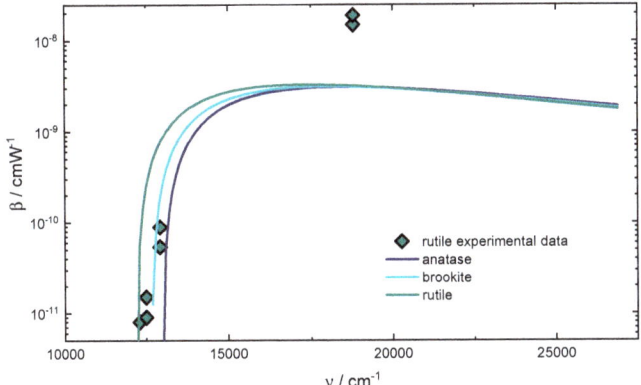

Figure 1. Two-photon absorption (TPA) coefficient as calculated from Equation (1) for crystalline TiO$_2$ modifications (solid lines; input data according to Table 1). Symbols: Experimental data reported in References [19,25] for rutile, as obtained from Z-scan techniques.

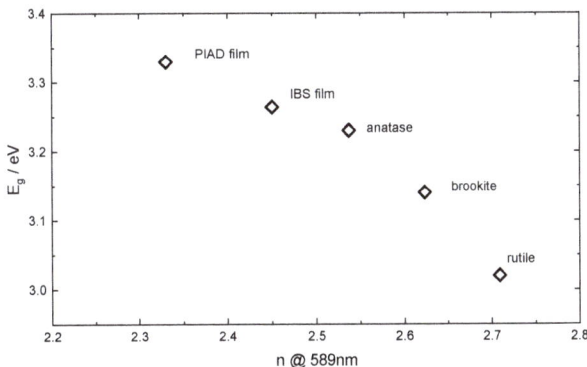

Figure 2. Correlation between optical gap and refractive index for the TiO$_2$ modifications from Table 1.

As previously mentioned, reported values for the nonlinear optical constants of TiO$_2$ are rare. Table 2 presents some published values of nonlinear refractive indices and absorption coefficients.

Note that the experimental β values published in Reference [25] are in rather good agreement with the theory (Figure 1), while those published in Reference [19] exceed the theoretically predicted values for a factor of approximately 5, such that experiment and theory indeed fall into the same order of magnitude anyway.

Table 2. Reported values for measured nonlinear TiO$_2$ optical constants. IAD denotes ion-assisted evaporation. C denotes the direction of the rutile optical axis, while E denotes the direction of the electric field vector in the light wave. IBS: Ion beam-sputtered.

Sample	Source	ν/cm^{-1}		n_2/cm^2 W^{-1}	ß/cm W^{-1}
Rutile single crystal	[19]	18,800	C‖E	-1.07×10^{-13}	$\approx 1.9 \times 10^{-8}$
			C⊥E	-1.02×10^{-13}	$\approx 1.5 \times 10^{-8}$
Anatase polycrystalline waveguide	[20]	6452		1.03×10^{-15}	-
				1.82×10^{-15}	-
Crystal	[21]	9434		$\approx 9.4 \times 10^{-15}$	-
IBS waveguide layer	[22]	6452		$\approx 3 \times 10^{-14}$	-
Rutile	[23]	9434		$\approx 2 \times 10^{-14}$	-
IAD thin film	[24]	19,800		$\approx 3 \times 10^{-12}$	-
Rutile	[25]	12,300	C‖E	-	$<1 \times 10^{-17}$
			C⊥E	-	8×10^{-12}
		12,500	C‖E	-	9×10^{-12}
			C⊥E	-	1.5×10^{-11}
		12,920	C‖E	-	8.9×10^{-11}
			C⊥E	-	5.4×10^{-11}
TiO$_2$ thin film	[26]	12,500		7.9×10^{-15}	-

As a second approach, we will make use of the beta-distributed oscillator (ß_do) model [16]. This is a semiempirical model primarily developed for fitting linear optical constants, but again, all of the input parameters can be fitted from linear optical spectra. The dielectric function ε in terms of the ß_do model is given by

$$\varepsilon(\nu) = [n(\nu) + ik(\nu)]^2 = 1 + \frac{J}{\pi} \frac{\sum_{s=1}^{M} w_s \left(\frac{1}{\nu_s - \nu - i\Gamma} + \frac{1}{\nu_s + \nu + i\Gamma} \right)}{\sum_{s=1}^{M} w_s} ; \qquad (2)$$
$$w_s = \left(\frac{s}{M+1} \right)^{A-1} \left(\frac{M+1-s}{M+1} \right)^{B-1}; s = 1,2,3,\ldots,M\,; A, B > 0$$
$$\nu_s = \nu_a + \frac{\nu_b - \nu_a}{M+1} s$$

while the real parameters J, Γ, A, B, ν_a, and ν_b are free parameters within the ß_do model [16], and M is the number of individual Lorentzian oscillators. Here, n and k are the linear refractive index and the extinction coefficient, correspondingly. Then, the third-order nonlinear susceptibility $\chi^{(3)}$ may be estimated by

$$\chi^{(3)} \approx J_3 g(\nu) f(\nu)^2 \left(\sum_{s=1}^{M} w_s \right)^{-3} ; \nu < \nu_a$$
$$f(\nu) = \sum_{s=1}^{M} w_s \left(\frac{1}{\nu_s - \nu - i\Gamma} + \frac{1}{\nu_s + \nu + i\Gamma} \right) , \qquad (3)$$
$$g(\nu) = \sum_{s=1}^{M} w_s \left(\frac{1}{\nu_s - 2\nu - i\Gamma} + \frac{1}{\nu_s + 2\nu + i\Gamma} \right)$$

which leads us to expressions for the nonlinear refractive index n_2 and the TPA coefficient β according to [27]

$$n_2(\nu) = \frac{3}{4} \frac{\mu_0 c}{(n^2 + k^2)} \left[\text{Re}\chi^{(3)} + \frac{k}{n} \text{Im}\chi^{(3)} \right]$$
$$\beta(\nu) = 3 \frac{\mu_0 \pi c \nu}{(n^2 + k^2)} \left[\text{Im}\chi^{(3)} - \frac{k}{n} \text{Re}\chi^{(3)} \right] , \qquad (4)$$

in which μ_0 represents the free space permeability. Note that the expressions written here are also valid in the case where there is still some linear absorption present in the TPA region [27]. In this approach, the parameter J_3 is estimated from the generalized millers rule according to References [14,16,28]. The gap values indicated in Table 1 for the PIAD and IBS samples correspond to hcv_a.

3. Experiment

3.1. PIAD Film Deposition

All samples were prepared in a Bühler Syrus LCIII deposition plant at Fraunhofer IOF using an electron-beam gun HPE6 and Ti_3O_5 as starter material. A target layer thickness of 200 nm was controlled by quartz crystal monitoring. For characterization, two different fused silica substrates (spectrophotometric characterization: Diameter 25 mm, thickness 1 mm; direct absorption measurement: Rectangular block $20 \times 20 \times 6$ mm^3) located in adjacent positions with identical radial positions in the rotating substrate holder, as well as silicon wafers, were used. During layer growth, additional energetic particle bombardment by means of a Bühler Advanced Plasma Source APS pro was accomplished. For this, two gas fluxes, Γ_1 and Γ_2 (for either argon or xenon as an inert gas), were used. In all experiments, the oxygen flow Γ_3 was 15 sccm, and the substrate temperature was around 110 °C. Additional main deposition parameters are summarized in Table 3.

Table 3. Main deposition parameters for the plasma ion-assisted evaporation (PIAD) preparation of the titanium dioxide single-layer coatings.

Sample	r/nms^{-1}	U_B/V	Inert Gas	Γ_1/sccm	Γ_2/sccm
1	0.1	120	Ar	6	6
2	0.5	160	Xe	2	2
3	0.1	160			

3.2. Energy-Dispersive X-Ray Spectroscopy (EDX)

EDX measurements were performed using a high-resolution scanning electron microscope (FE-SEM Sigma, Carl Zeiss Microscopy GmbH, Oberkochen, Germany). Spectra were analyzed using INCA Software (INCA Energy 250, INCA Oxford Instruments GmbH, Wiesbaden, Germany). EDX measurements were performed with samples deposited onto silicon substrates. As the EDX method is not particularly surface-sensitive, the detection volume is dependent on the acceleration voltage and includes the film as well as parts of the substrate. In order to eliminate the substrate response, the film atomic composition was estimated from a combined elaboration of two EDX spectra obtained from different acceleration voltages (10 kV and 16 kV) for each sample.

3.3. Spectrophotometric Characterization

The transmittance (T) and reflectance (R) of all samples were measured in the spectral range of 320–2000 nm at near-normal incidence as well as at an incidence angle of 60° (in s- and p-polarization) in a Perkin Elmer Lambda 900 scanning spectrophotometer using the VN measurement technique [29]. From these spectra, film thickness d as well as optical constants n and k were obtained as a result of spectra fits in terms of Equation (2) (using a Matlab environment). In all spectra fits, the number of individual Lorentzian oscillators M in Equation (2) was set to $M = 10,000$, which has been proven to model a smooth function [16].

3.4. Direct Absorption Measurements

Light absorption measurements in TiO_2 films were measured directly using a laser-induced deflection (LID) technique [30,31]. This technique belongs to an ensemble of photothermal techniques with a pump–probe configuration. When the pump laser hits the sample under investigation, the absorbed pump laser power forms a temperature profile (Figure 3, left). The latter is turned into a refractive index profile (=the thermal lens) by both the thermal expansion and the temperature-dependent refractive index. The refractive index gradient accounts for a deflection of the probe beams (from the same laser source) that is proportional to the absorbed pump laser power.

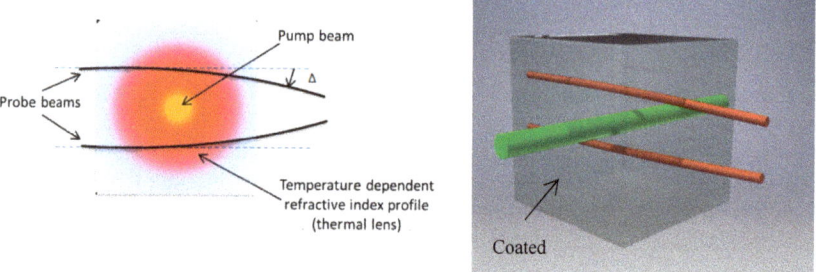

Figure 3. General scheme for the laser-induced deflection (LID) photothermal technique (**left**) and the measurement concept for rectangular substrate geometries (**right**).

Figure 3 (right) shows the applied measurement concept for the investigated rectangular substrate geometry (20 × 20 × 6 mm^3) with one coated surface. Two-probe beams above/beneath the irradiated spot utilize the probe beam deflection perpendicular to the pump beam direction. To measure coatings, the probe beams pass the sample close to the coated surface. In the case of transparent coatings, the probe beam deflection always comprises both the coating and substrate absorption. In order to distinguish both absorption contributions, an uncoated reference substrate of the same geometry/material is measured additionally, and the difference in the deflection signals is assigned to the coating absorption.

Calibration of the measurement setup is required to obtain absolute absorptance data from the deflection signals. For the LID technique, electrical calibration is applied, i.e., the thermal lens is generated by particular electric heaters. In the case of coating/surface absorption, small surface-mounted device (SMD) elements—fixed onto a very thin copper plate (thickness ~200 μm)—are placed centrally onto the surface of a reference substrate (of the same geometry and material) [32]. The copper plate allows for the required high thermal conduction to the sample. The validity of this calibration approach has been verified through separate measurements of reflectance, transmittance, absorptance, and scattering for different materials and coatings. The results of these energy balance measurements confirmed that in terms of measurement accuracy, a value of 1 was obtained in each of the investigations [33]. The calibration procedure itself is composed of measuring the probe beam deflection as a function of the electric power. Plotting the deflection signals versus electrical power (Figure 4) gives a linear function that spans several orders of magnitude for electric power, and the calibration coefficient F_{CAL} is defined by the slope of the linear function (including the zero-point, i.e., no electrical power means no probe beam deflection) [31]. From the LID deflection signal I_{LID} (for the sample under investigation), the corresponding mean pump laser power P_L, and the calibration coefficient F_{CAL}, the coating absorptance (defined as the ratio of the absorbed and incident light intensity) is calculated by

$$\text{absorptance} = \frac{I_{LID}}{F_{CAL} P_L}. \tag{5}$$

Laser irradiation at around 800 nm was realized by two different laser sources. For low-intensity measurements, an 808-nm continuous-wave semiconductor laser (HangZhou Naku Technology Co., Ltd.) with a maximum output power of 10 W was applied. The laser beam was shaped to a spatial profile of about 2 × 2 mm^2 on the sample. For elevated laser intensities in the GW/cm^2 range, an 800-nm Femtosecond laser (Astrella-V-F-1k, Coherent Inc.) with a pulse duration of 82 fs, a repetition rate of 1 kHz, and an average power of up to 2.1 W was used. In order to vary the laser intensity and maintain the laser pulse duration, a combination of a thin-film polarizer and a polarizing beam splitter was placed into the beam path. A telescope was used to shape the Gaussian laser output beam to a $1/e^2$ Diameter of 5.4 mm on the sample under investigation.

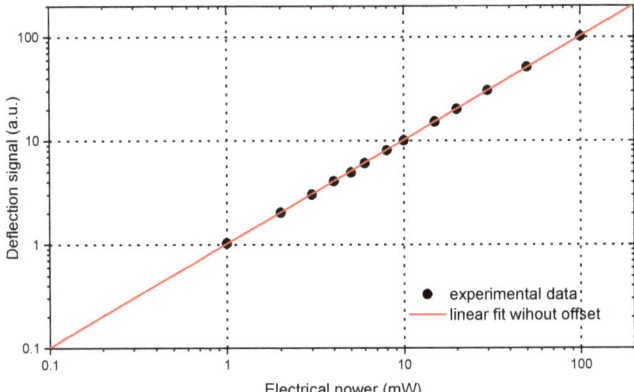

Figure 4. Measured deflection signal as a function of the electrical power (dots) and the linear fit without offset (red line) used for a determination of the calibration coefficient.

For a wavelength of 800 nm, the optical thickness of a 200-nm-thick TiO_2 film is not equal to $\lambda/4$ or multiples of $\lambda/4$. Therefore, laser beam reflection at the interface of the air/TiO_2 film was not negligible. The calculated reflectance amounts (up to about 15%) were taken into account for the determination of the average laser power P_L inside the TiO_2 film.

4. Results

4.1. Linear Optical Constants from Spectrophotometry and Their Parametrization

The linear optical constants of the titanium dioxide samples were fitted by means of Equation (2). In Figure 5, the measured and modeled transmittance and reflectance for sample 3 at near-normal incidence and at 60° for s- and p-polarization and the corresponding optical constants are shown. The complete set of calculated model parameters is summarized in Table 4.

Table 4. Here, beta-distributed oscillator (ß_do) parameters of all samples obtained from the spectra fitting.

Sample	M	v_a/cm^{-1}	v_b/cm^{-1}	Γ/cm^{-1}	A	B	J/cm^{-1}	d/nm	J_3/V^{-2} cm^5
1	10,000	26,690	106,690	40	3.15	7.35	291,960	201.5	0.001679
2	10,000	26,975	106,792	40	3.14	7.32	287,543	194.6	0.001570
3	10,000	26,850	106,850	50	3.07	7.04	291,600	195.5	0.001654

In order to verify the film elementary composition, EDX measurements were performed. Uncorrected (raw) data on the elementary composition contained a significant substrate contribution, which increased with increasing acceleration voltage. In order to eliminate the substrate contribution, each sample was measured with two acceleration voltages, and the (corrected) elementary composition of the film was estimated using Equation (6):

$$N_{corrected,Ti} = \frac{N_{raw,Si,16}N_{raw,Ti,10} - N_{raw,Si,10}N_{raw,Ti,16}}{N_{raw,Si,16} - N_{raw,Si,10}}$$
$$N_{corrected,ng} = \frac{1}{2}\left[\frac{N_{raw,ng,16}}{N_{raw,Ti,16}} + \frac{N_{raw,ng,10}}{N_{raw,Ti,10}}\right]N_{corrected,Ti} \quad . \tag{6}$$
$$N_{corrected,O} = 1 - N_{corrected,Ti} - N_{corrected,ng}$$

Generally, N denotes the atomic concentration, whereas "raw" and "corrected" subscript indicate raw and corrected values, respectively. Accordingly, "10" or "16" subscripts indicate the raw values obtained at 10 and 16 kV of acceleration voltage, respectively. Here, ng stands for the noble gas used, i.e., argon or xenon.

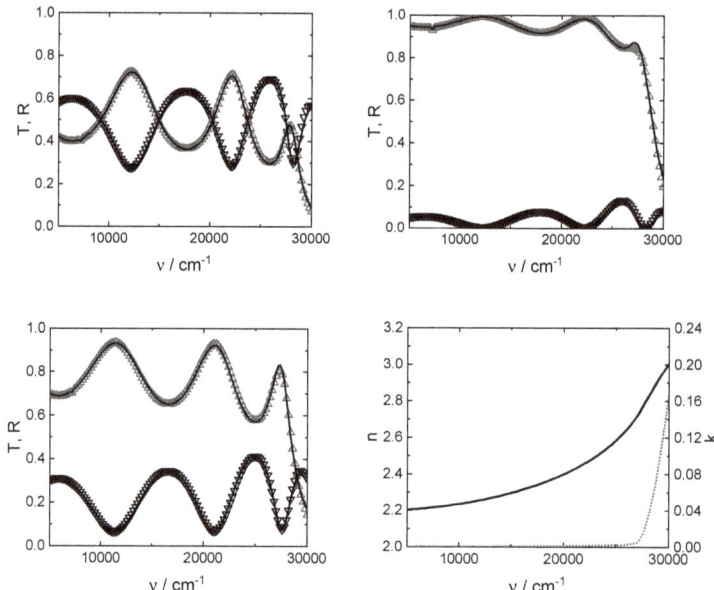

Figure 5. Modeled (solid line) and measured transmittance (up triangles) and reflectance (down triangles) at 60° for s- (**top left**) and p-polarizations (**top right**) and at near-normal incidence (**bottom left**) for sample 3 and modeled optical constants (**bottom right**, refractive index: Solid line, left axes; extinction coefficient: Dotted line, right axes).

The final calculated atomic concentrations are summarized in Table 5.

Table 5. Energy-dispersive X-ray spectroscopy (EDX) results and linear refractive index of the PIAD films.

Sample	$N_{corrected,Ti}$ at %	$N_{corrected,O}$ at %	$N_{corrected,Ar}$ at %	$N_{corrected,Xe}$ at %	n@800 nm (12,500 cm^{-1})
1	32	64	4	-	2.269
2	31	66	-	3	2.253
3	32	65	-	3	2.261

Note that sample 1 had the highest refractive index, such that we expected the largest density for that sample. This was in agreement with the EDX results, indicating a correct stoichiometric relation between titanium and oxygen atoms. Samples 2 and 3 showed some more oxygen in the EDX analysis, indicating a somewhat porous layer structure with some incorporated water molecules. Correspondingly, the refractive index turned out to be smaller. Noble gas impurities from the used inert gas of the plasma source (Table 3) were of the order of 3 to 4 at % in all samples.

4.2. LID: Absorption Properties of the Films

Figures 6–8 give the film absorptance as a function of the laser intensity for the three investigated TiO$_2$ films. Table 6 summarizes the linear absorptance and the TPA coefficients.

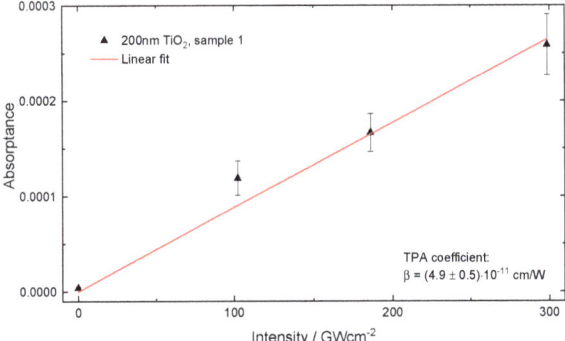

Figure 6. Absorptance versus laser intensity and the TPA coefficient for sample 1.

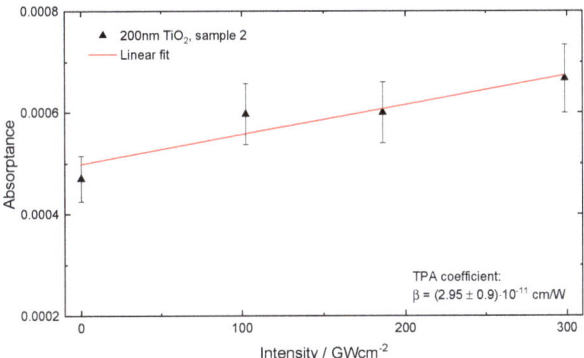

Figure 7. Absorptance versus laser intensity and the TPA coefficient for sample 2.

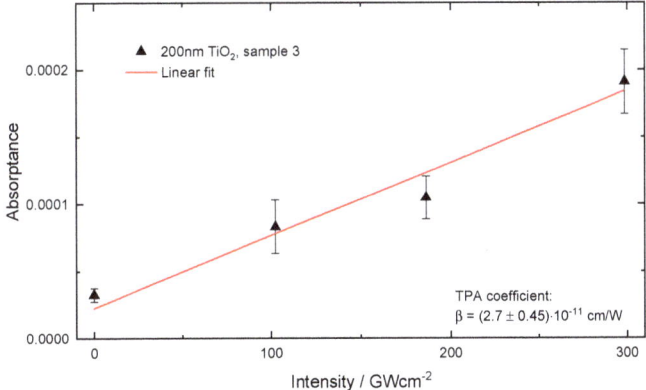

Figure 8. Absorptance versus laser intensity and the TPA coefficient for sample 3.

Table 6. LID results.

TiO$_2$ Film	Linear Absorptance at 808 nm (ppm)	TPA Coefficient at 800 nm (12,500 cm^{-1}) (10^{-11} cm/W)
1	(3.9 ± 1.3)	(4.9 ± 0.5)
2	(470 ± 50)	(3.0 ± 1.0)
3	(32 ± 6)	(2.7 ± 0.5)

5. Discussion

5.1. Film Absorption

Linear absorption in TiO$_2$ films at 800 nm has been shown to be as low as a few ppm, but varies over two orders of magnitude. No correlation was found between linear and nonlinear absorption properties. The observed TPA coefficients were less than one order of magnitude higher than previously published TPA coefficients in the range of 0.9–1.5 × 10^{-11} cm/W (measured at 800 nm in bulk rutile (TiO$_2$)) [25]. However, this finding is in agreement with earlier experimental results for fluoride thin films at 193 nm [34]. Here, it was demonstrated that the TPA coefficients of thin films were significantly larger than those obtained for the corresponding bulk materials.

On the other hand, the TPA coefficients observed in this study were about three orders of magnitude smaller than the values published in Reference [19] (compare Table 2), which was obviously a result of the different wavenumbers. Indeed, the photon energy in our study was still slightly below the value of $E_g/2$ (which we will further call the TPA threshold energy), while in Reference [19], the photon energy was well above that value. In order to visualize that, Figure 9 presents the measured TPA coefficients together with those calculated according to both the Sheik–Bahae (Equation (1), solid lines) and ß_do (dashed lines) models. Note that both models predicted an increase in β by almost three orders of magnitude when the wavenumber was changed from 12,500 to 18,800 cm^{-1}. This way, both models reproduced the dynamic range observed in the measured values. What is particularly remarkable is the good mutual agreement between both model approaches in the mentioned spectral region: The models delivered remarkably different results only when the photon energy came close to the single photon absorption edge at wavenumbers around 26,500 cm^{-1}. For reference purposes, we parametrized the dielectric function of the IBS sample specified in References [17,18] in terms of the ß_do model and included corresponding estimations of the TPA coefficient with Figure 9: Similarly to what is shown in Figure 1, the Sheik–Bahae model predicted a wavenumber shift of the spectral features without significant changes in the maximum absorption. On the contrary, note that in the ß_do model, the larger density of the IBS layer resulted in an increase in β, so that the calculated TPA coefficient came closer to the rutile value from Reference [19]. Again, for reference purposes, we included the rutile TPA simulation in terms of Equation (1) with Figure 9.

Note that the wavenumbers around 12,500 cm^{-1} fell close to the TPA threshold wavenumbers in TiO$_2$. In the case of the rutile data from Reference [25], the Sheik–Bahae model (dark cyan line) provided a good theoretical reproduction of the TPA coefficients, because the excitation wavenumbers (Table 2) were still higher than the corresponding TPA threshold wavenumber (12,177 cm^{-1}, compare Table 1), which was required for calculation in terms of Equation (1). However, the measured PIAD data could not be reproduced in terms of the Sheik–Bahae model (solid black line) with the parameters given in Table 1. The reason is that in Equation (1), the excitation wavenumber should exceed the TPA threshold, which corresponds to 13,345–13,490 cm^{-1} for the PIAD samples (Tables 1 and 4). This was not achieved in our measurements. Clearly, the PIAD films were essentially amorphous, such that the optical gap (and correspondingly the TPA thresholds) did not represent "hard" threshold energies. Instead, band tailing allowed for absorption even when the photon energy was smaller than the corresponding "threshold". Note that in this connection, the ß_do model predicted a certain TPA even below the thresholds, which turned out to be less than one order of magnitude smaller than the measured absorption. Clearly, films produced by evaporation are usually highly defective and maybe somewhat contaminated, and it is therefore not so surprising that the measured absorption values somewhat exceeded the modeled ones.

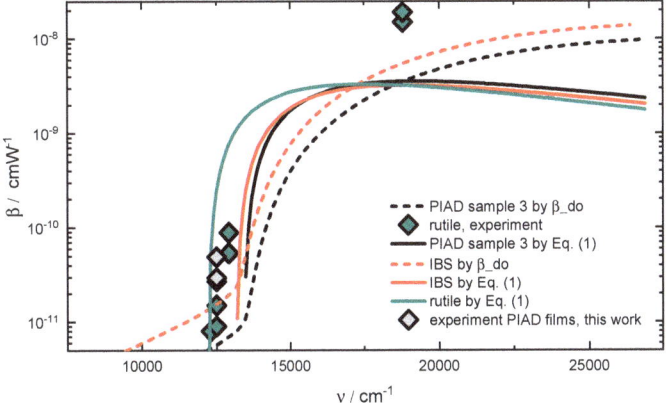

Figure 9. Measured TPA coefficients from rutile and PIAD TiO$_2$ compared to simulations in terms of Equations (1), (3), and (4).

5.2. Considerations Regarding Nonlinear Refraction

It should be mentioned that both model approaches allowed for estimating the nonlinear refractive index as well, and such like calculations could be compared to the reported experimental data collected in Table 2. This is shown in Figure 10 (excluding the extraordinarily large value from Reference [24]). The predictions of the Sheik–Bahae model (assuming the input data from Table 1) almost coincided with the chosen ordinate scaling and were in reasonable agreement with the experimental data from References [20–23,26]. The ß_do model delivered results in the infrared that practically coincided with the Sheik–Bahae model predictions up to the TPA threshold. At higher wavenumbers, both approaches delivered divergent results. Note that none of the models was able to reproduce the strongly negative n_2 value from Reference [19], but the Sheik–Bahae model came at least close to this value. This is not astonishing, because in the ß_do approach (Equations (3) and (4)), only the TPA resonant contribution to n_2 is taken into account, while nonresonant contributions are not considered at all.

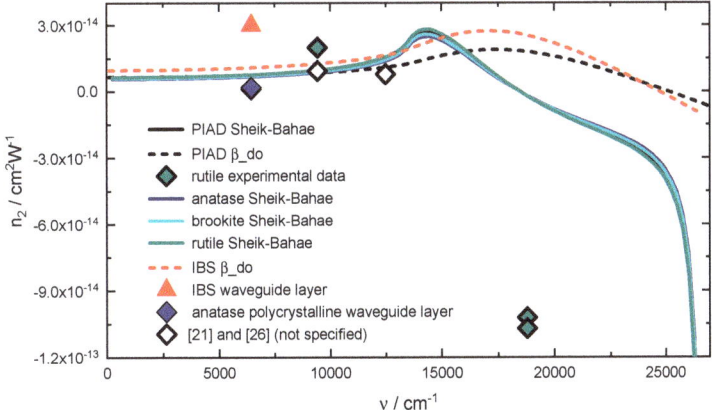

Figure 10. Measured nonlinear refractive indices from different TiO$_2$ modifications compared to simulations in terms of the Sheik–Bahae approach as well as Equations (3) and (4).

The relative stability of the Sheik–Bahae approach with respect to correlated changes in the optical gap and the refractive index makes it difficult to understand density-related differences in the nonlinear response of different TiO$_2$ modifications. Nevertheless, a density dependence of n_2 and β is physically reasonable and expected. It seems to be an advantage of the ß_do model that differences in the density are explicitly taken into account in terms of the model parameter J; consequently, the modeled n_2 data for the IBS film is larger than those of the PIAD film. This way, the ß_do model might provide access to explaining the significant scatter in measured nonlinear constants, as has been reported in the different studies cited. For future practical modeling, it might therefore be useful to merge both models in a suitable way.

5.3. Considerations Regarding the Effect of the Nonlinear Film Properties on Elastic Scattering

Albeit not the main scope of this paper, we would like to briefly consider the role of light scattering and the potential impact of nonlinear effects on light scattering properties. We focus here on elastic scattering processes, because with conventional laser coatings, elastic scattering is four orders of magnitude stronger than inelastic scattering. Nevertheless, one must keep in mind that the excitation of nonlinear optical processes may also result in an increase in inelastic scatter contributions, but this discussion is outside the scope of this paper.

For most interference coatings, the interface roughness is the dominating source of light scattering. The angle-resolved scattering (ARS) can be calculated using multilayer vector perturbation (VPT) theories (given in detail, for instance, in References [35–37] and summarized briefly in Reference [38]). The most relevant factors influencing light scattering properties are (i) interface roughness, (ii) the cross-correlation properties of the roughness of different interfaces, (iii) the field strengths at the interfaces, and (iv) the interference properties not only for the specular fields but also for the scattered fields (the latter of course being linked to wavelengths and incident angles). Defects and contamination on the coatings can be considered to be additional sources of light scattering, which we excluded from this discussion.

As a result of VPT, the angle-resolved scattering distribution (the scattered intensity) of a multilayer can be calculated as

$$ARS(\theta_s) = \frac{16\pi^2}{\lambda^4} \sum_{i=0}^{Z} \sum_{j=0}^{Z} F_i F_j^* PSD_{ij}(f), \qquad (7)$$

where Z is the number of layers (of course $Z = 1$ for a single film), and F_i and F_j are optical factors containing information about the optical properties of the perfectly smooth multilayer (design, dielectric functions, etc.) and the conditions of illumination and detection (angles, polarization, etc.). The roughness factor PSD_{ij} comprises the power spectral density functions of all interfaces (for $i = j$) and their cross-correlation properties (for $i \neq j$). The total scatter can be calculated by integrating Equation (7).

It is important to go back into the full treatment of the VPTs to assess the influence of nonlinear effects. The main approach of VPT is that the unperturbed specular fields are used to calculate the fields at the interfaces. Depending on the roughness structure, these fields drive surface currents, which produce scattered waves. These scattered waves then propagate through the coating and interfere. Therefore, we expect an influence of nonlinear effects on the scattering properties in two ways: (i) the field-induced change of the dielectric function leads to a modification of the field distribution inside the coating and hence modified fields at the interfaces, and (ii) the modified dielectric properties change the propagation and interference properties of the scattered waves. It has been demonstrated that even small changes in the field distribution, e.g., caused by small wavelength shifts, can easily lead to an enhancement of the scatter losses by an order of magnitude [38]. Simply put, if for whatever reason the observed specular field deviates from the expected values by several percentage points, a substantial change in the scatter losses can be expected.

Although not trivial, we believe that taking nonlinear effects into account in scatter modeling is straightforward. One precondition is that the scattering effects are still so weak that they do not influence the determination of the dielectric function, including nonlinear effects using the approaches described in this paper. We should then be able to use these new parameters to calculate the scattered fields. Further studies will show if this pragmatic approach is justified. We also believe that for interference coatings, once we have taken nonlinear effects into account in the design, we are also able to bring the light scattering properties back to the level we would expect for a similar coating designed for low fluences, or we even have the possibility of making use of nonlinear effects for scatter loss reduction.

6. Conclusions

We presented experimental results on the optical properties of PIAD titanium dioxide thin films, with a focus on their absorption behavior. In addition to a rather conventional linear optical characterization, TPA coefficients at a wavelength of 800 nm were obtained from laser-induced deflection measurements and were compared to the values reported so far for rutile crystals in terms of a Z-scan technique. It turns out that the TPA coefficients obtained from our LID measurements were very consistent with what has been reported in the literature in terms of manageable models describing the TPA coefficient, such as the Sheik–Bahae model and the ß_do approach.

We provided a further discussion on the accompanying changes in the refractive index in terms of nonlinear refraction and discussed their possible impact on the elastic light scattering characteristics of dielectric coatings at high light intensities.

Author Contributions: Conceptualization, O.S.; formal analysis, O.S., S.W., S.S., and C.M.; funding acquisition, S.W.; methodology, O.S., S.W., and C.M.; project administration, S.W.; software, S.W.; validation, O.S., S.W., S.S., and C.M.; writing—Original draft, O.S.; writing—Review and editing, O.S., S.W., S.S., and C.M. All authors have read and agreed to the published version of the manuscript.

Funding: This research was funded by the Bundesministerium für Bildung und Forschung, grant number 13N10459, and by the European Regional Development Fund, grant number 2016 FE 9045.

Acknowledgments: The authors are grateful to Tina Seifert (Fraunhofer IOF) for performing the EDX measurements. We also thank Tsvetanka Babeva for the invitation to submit this study to a special issue of "Coatings" as a feature paper.

Conflicts of Interest: The authors declare no conflicts of interest. The funders had no role in the design of the study; in the collection, analyses, or interpretation of data; in the writing of the manuscript; or in the decision to publish the results.

References

1. Born, M.; Wolf, E. *Principles of Optics*; Pergamon Press: Oxford, UK, 1968.
2. Furman, S.A.; Tikhonravov, A.V. *Basics of Optics of Multilayer Systems*; Editions Frontières: Paris, France, 1992.
3. Stenzel, O. *The Physics of Thin Film Optical Spectra: An Introduction*, 2nd ed.; Springer: Berlin, Germany, 2016.
4. Macleod, H.A. *Thin-Film Optical Filters*, 4th ed.; CRC Press: Boca Raton, FL, USA, 2010.
5. Thelen, A. *Design of Optical Interference Coatings*; McGraw-Hill: New York, NY, USA, 1989.
6. Shen, Y.R. *The Principles of Nonlinear Optics*; John Wiley & Sons Ltd.: Hoboken, NJ, USA, 1984.
7. Mourou, G.; Mironov, S.; Khazanov, E.; Sergeev, A. Single cycle thin film compressor opening the door to Zeptosecond-Exawatt physics. *Eur. Phys. J. Spec. Top.* **2014**, *223*, 1181–1188. [CrossRef]
8. Razskazovskaya, O.; Luu, T.T.; Trubetskov, M.; Goulielmakis, E.; Pervak, V. Nonlinear Behavior and Damage of Dispersive Multilayer Optical Coatings Induced by Two-Photon Absorption. *Proc. SPIE* **2014**, *9237*, 92370L1–92370L8.
9. Amotchkina, T.; Trubetskov, M.; Fedulova, E.; Fritsch, K.; Pronin, O.; Krausz, F.; Pervak, V. Characterization of Nonlinear Effects in Edge Filters. In Proceedings of the Optical Interference Coatings (OIC), Tucson, AZ, USA, 19–24 June 2016. Paper ThD.3.
10. Stenzel, O.; Wilbrandt, S. Theoretical study of multilayer coating reflection taking into account third-order optical nonlinearities. *Appl. Opt.* **2018**, *57*, 8640–8647. [CrossRef] [PubMed]

11. Tichá, H.; Tichý, L. Semiempirical relation between non-linear susceptibility (refractive index), linear refractive index and optical gap and its application to amorphous chalcogenides. *J. Optoelectron. Adv. Mater.* **2002**, *4*, 381–386.
12. Fournier, J.; Snitzer, E. The nonlinear refractive index of glass. *IEEE J. Quantum Electron.* **1974**, *10*, 473–475. [CrossRef]
13. Boling, N.L.; Glass, A.J.; Owyoung, A. Empirical Relationships for Predicting Nonlinear Refractive Index Changes in Optical Solids. *IEEE J. Quantum Electron.* **1968**, *14*, 601–608. [CrossRef]
14. Stenzel, O. Simplified expression for estimating the nonlinear refractive index of typical optical coating materials. *Appl. Opt.* **2017**, *56*, C21–C23. [CrossRef]
15. Sheik-Bahae, M.; Hutchings, D.C.; Hagan, D.J.; Van Stryland, E.W. Dispersion of Bound Electronic Nonlinear Refraction in Solids. *IEEE J. Quantum Electron.* **1991**, *27*, 1296–1309. [CrossRef]
16. Stenzel, O.; Wilbrandt, S. Beta-distributed oscillator model as an empirical extension to the Lorentzian oscillator model: Physical interpretation of the β_do model parameters. *Appl. Opt.* **2019**, *58*, 9318–9325. [CrossRef]
17. Stenzel, O.; Wilbrandt, S.; Kaiser, N.; Schmitz, C.; Turowski, M.; Ristau, D.; Awakowicz, P.; Brinkmann, R.P.; Musch, T.; Rolfes, I.; et al. Plasma and optical thin film technologies. *Proc. SPIE* **2011**, *8168*, 81680L.
18. Landmann, M.; Köhler, T.; Köppen, S.; Rauls, E.; Frauenheim, T.; Schmidt, W.G. Fingerprints of order and disorder in the electronic and optical properties of crystalline and amorphous TiO_2. *Phys. Rev. B* **2012**, *86*, 064201. [CrossRef]
19. Watanabe, Y.; Ohnishi, M.; Tsuchiya, T. Measurement of nonlinear absorption and refraction in titanium dioxide single crystal by using a phase distortion method. *Appl. Phys. Lett.* **1995**, *66*, 3431–3432. [CrossRef]
20. Shtyrkova, K. Characterization of Third Order Nonlinearities in TiO_2 Waveguides at 1550 nm. Master's Thesis, Massachusetts Institute of Technology, Cambridge, MA, USA, 2013.
21. Adair, R.; Chase, L.L.; Payne, S.A. Nonlinear refractive index of optical crystals. *Phys. Rev. B* **1989**, *39*, 3337–3350. [CrossRef]
22. Guan, X.; Hu, H.; Oxenløwe, L.K.; Frandsen, L.H. Compact titanium dioxide waveguides with high nonlinearity at telecommunication wavelengths. *Opt. Express* **2018**, *26*, 1055–1063. [CrossRef] [PubMed]
23. Friberg, S.R.; Smith, P.W. Nonlinear Optical Glasses for Ultrafast Optical Switches. *IEEE J. Quantum Electron.* **1987**, *23*, 2089–2094. [CrossRef]
24. Rigneault, H.; Flory, F.; Monneret, S. Nonlinear totally reflecting prism coupler: Thermomechanic effects and intensity-dependent refractive index of thin films. *Appl. Opt.* **1995**, *34*, 4358–4369. [CrossRef] [PubMed]
25. Evans, C.C.; Bradley, J.D.B.; Martí-Panameño, E.A.; Mazur, E. Mixed two- and three-photon absorption in bulk rutile (TiO_2) around 800 nm. *Opt. Exp.* **2012**, *20*, 3118–3128. [CrossRef]
26. Evans, C.C.; Bradley, J.D.B.; Parsy, F.; Phillips, K.C.; Senaratne, R.; Martí-Panameño, E.A.; Mazur, E. Thermally managed Z-scan measurements of titanium dioxide thin films. presented at Photonics West, San Francisco, CA, USA, 27 January 2011.
27. del Coso, R.; Solis, J. Relation between nonlinear refractive index and third-order susceptibility in absorbing media. *J. Opt. Soc. Am. B* **2004**, *21*, 640–644. [CrossRef]
28. Wang, C.C. Empirical Relation between the Linear and the third-order Nonlinear Optical Susceptibilities. *Phys. Rev. B* **1970**, *2*, 2045–2048. [CrossRef]
29. Stenzel, O. *Optical Coatings: Material Aspects in Theory and Practice*; Springer: Berlin, Germany, 2014; pp. 117–127.
30. Guntau, M.; Triebel, W. A Novel method to measure bulk absorption in optically transparent materials. *Rev. Sci. Instr.* **2000**, *71*, 2279–2282. [CrossRef]
31. Bublitz, S.; Mühlig, C. Absolute Absorption Measurements in Optical Coatings by Laser Induced Deflection. *Coatings* **2019**, *9*, 473. [CrossRef]
32. Mühlig, C.; Triebel, W.; Kufert, S.; Bublitz, S. Characterization of low losses in optical thin films and materials. *Appl. Opt.* **2008**, *47*, C135–C142. [CrossRef] [PubMed]
33. Mühlig, C.; Bublitz, S. Sensitive and absolute absorption measurements in optical materials and coatings by laser induced deflection (LID) technique. *Opt. Eng.* **2012**, *51*, 121812. [CrossRef]
34. Mühlig, C.; Bublitz, S.; Kufert, S. Nonlinear absorption in single LaF_3 and MgF_2 layers at 193 nm measured by surface sensitive laser induced deflection technique. *Appl. Opt.* **2009**, *48*, 6781–6787. [CrossRef] [PubMed]

35. Elson, J.; Rahn, J.; Bennett, J. Light scattering from multilayer optics: Comparison of theory and experiment. *Appl. Opt.* **1980**, *19*, 669–679. [CrossRef]
36. Bousquet, P.; Flory, F.; Roche, P. Scattering from multilayer thin films: Theory and experiment. *J. Opt. Soc. Am.* **1981**, *71*, 1115–1123. [CrossRef]
37. Amra, C.; Grezes-Besset, C.; Roche, P.; Pelletier, E. Description of a scattering apparatus: Application to the problems of characterization of opaque surfaces. *Appl. Opt.* **1989**, *28*, 2723–2730. [CrossRef]
38. Schröder, S.; Trost, M.; Herffurth, T.; von Finck, A.; Duparré, A. Light scattering of interference coatings from the IR to the EUV spectral regions. *Adv. Opt. Technol.* **2013**, *3*, 113–120. [CrossRef]

© 2020 by the authors. Licensee MDPI, Basel, Switzerland. This article is an open access article distributed under the terms and conditions of the Creative Commons Attribution (CC BY) license (http://creativecommons.org/licenses/by/4.0/).

Article

Atomic Layer-Deposited Al-Doped ZnO Thin Films for Display Applications

Dimitre Dimitrov [1,2,*], Che-Liang Tsai [3], Stefan Petrov [4], Vera Marinova [2,4,*], Dimitrina Petrova [2,5], Blagovest Napoleonov [2], Blagoy Blagoev [1], Velichka Strijkova [2], Ken Yuh Hsu [3] and Shiuan Huei Lin [4]

1. Institute of Solid State Physics, Bulgarian Academy of Sciences, 1784 Sofia, Bulgaria; blago_sb@yahoo.com
2. Institute of Optical Materials and Technologies, Bulgarian Academy of Sciences, 1113 Sofia, Bulgaria; d_kerina@abv.bg (D.P.); blgv@abv.bg (B.N.); vily@iomt.bas.bg (V.S.)
3. Department of Photonics, National Chiao Tung University, Hsinchu 30010, Taiwan; Patrick.Tsai@crystalvue.com.tw (C.-L.T.); ken@cc.nctu.edu.tw (K.Y.H.)
4. Department of Electrophysics, National Chiao Tung University, Hsinchu 30010, Taiwan; st5pob@gmail.com (S.P.); lin@cc.nctu.edu.tw (S.H.L.)
5. Faculty of Engineering, South-West University "Neofit Rilski", 2700 Blagoevgrad, Bulgaria
* Correspondence: dzdimitrov@issp.bas.bg (D.D.); vmarinova@iomt.bas.bg (V.M.)

Received: 30 April 2020; Accepted: 25 May 2020; Published: 31 May 2020

Abstract: The integration of high uniformity, conformal and compact transparent conductive layers into next generation indium tin oxide (ITO)-free optoelectronics, including wearable and bendable structures, is a huge challenge. In this study, we demonstrate the transparent and conductive functionality of aluminum-doped zinc oxide (AZO) thin films deposited on glass as well as on polyethylene terephthalate (PET) flexible substrates by using an atomic layer deposition (ALD) technique. AZO thin films possess high optical transmittance at visible and near-infrared spectral range and electrical properties competitive to commercial ITO layers. AZO layers deposited on flexible PET substrates demonstrate stable sheet resistance over 1000 bending cycles. Based on the performed optical and electrical characterizations, several applications of ALD AZO as transparent conductive layers are shown—AZO/glass-supported liquid crystal (LC) display and AZO/PET-based flexible polymer-dispersed liquid crystal (PDLC) devices.

Keywords: Al-doped ZnO; ALD technique; transparent conductive layers; LC display; flexible PDLC devices

1. Introduction

Transparent conducting materials (TCMs) possess simultaneously high electrical conductivity and optical transparency (i.e., effective transmission of light in the visible spectrum), with these particular characteristics being achieved on processing the materials as thin films on transparent substrates. This specific combination of properties makes TCMs very interesting both from a fundamental viewpoint and for a large variety of applications. The principal factor that has stimulated the research on TCM syntheses and their processing is definitely the development of optoelectronic materials and devices in which the principles of actuation involve an application of electric current or voltage to control the emission or passage of light, the most notable examples being display devices [1,2]. Moreover, other applications that require TCM, such as smart windows (based on electrochromic or polymer-dispersed liquid crystalline materials) and photovoltaic systems, have attained enormous relevance in the present context of environmentally important energy efficiency and clean energies, further prompting the scientific and technical developments in TCMs, as evidenced in the intensive and continuing research in this field [3].

The most studied and practically used TCMs are oxides, referred to as transparent conducting oxides (TCOs). TCOs are usually made of high band gap (> 3 eV) oxides which are intrinsically or extrinsically doped to reach a very low resistivity (~10^{-4} $\Omega\cdot$cm). The basic TCO materials include indium oxide (In_2O_3), tin oxide (SnO_2), and zinc oxide (ZnO). They can be degenerately n-type doped with tin (In_2O_3:Sn, also known as ITO), fluorine (SnO_2:F) or Al (ZnO:Al), as popular examples. The usage of TCO films depends on the application. In liquid crystal displays (LCDs), light-emitting diodes (LEDs), or transparent displays, these films are used as electrodes, while they are used as touch sensors for resistive and capacitive touch panels. Transparent electrodes are essential in high-impact technological areas such as photovoltaics, flat panel displays and touch screens, as well as in emerging areas such as smart sensors or organic electronics (organic light-emitting devices (OLEDs), organic photovoltaics) as well as transparent field-effect transistors (FETs) [4]. The most relevant properties for all these applications are their high conductivity combined with high transparency in the visible spectral region. Furthermore, key TCO performance factors, depending on the particular application, are their high chemical and thermal stability or the possibility of tuning their work function. Especially important for flexible electronics are the mechanical properties, high stretchability and bendability, and low-contact resistance with organic materials. Transparent conducting films are estimated to reach a market of value of US$ 8.46 billion by 2026 [5]; intensive research is therefore important to discover superior materials, new substrates, and new ways to enhance light transmission, to increase the electric conductivity, to add flexibility, and to decrease costs.

ITO is the most widely used material for the fabrication of TCOs; however, ITO has several critical shortcomings such as generally high processing cost and unsuitability for flexible devices. In particular, for large-area touch screens, the resistivity is too high for the rapid touch sensing response; in addition, ITO is brittle and therefore inadequate for applications in flexible electronics such as flexible touch screen displays and solar cells. Aluminum-doped zinc oxide (AZO) is an affordable, non-toxic and robust transparent conductive oxide (TCO) [6]. AZO films have high transmission in the visible region and useable transmission to IR wavelengths as long as ~12 μm. In contrast, the more commonly known TCO, ITO, reflects IR at wavelengths longer than ~2 μm. IR transmission is very important because increasing the long-wavelength response is an approach to enhance the efficiency of some solar devices [7]. The higher stability of AZO in reducing atmospheres may also be an advantage for future applications [8]. A substantial cost saving is possible with AZO materials compared to ITO and other TCOs. Due to these practical advantages, AZO films are considered as ideal replacements for ITO films in applications such as transparent electrodes for solar cells, flat panel displays, LCD electrodes, touch panel transparent contacts and IR windows [9–11]. AZO thin films can be deposited by several techniques such as sol–gel [12], chemical spray [13], thermal evaporation [14], pulsed laser deposition [15], DC and RF magnetron sputtering [16], reactive mid-frequency magnetron sputtering using dual magnetrons [17] and atomic layer deposition (ALD) [18]. The desired properties of a good TCO—transparency, conductivity, and surface texture—depend significantly on the preparation technique and the growth parameters.

ALD is a growth technique that has recently become very popular since it provides uniform and conformal coverage and control of the thin film by atomic layer precision [19]. Although the growth rate of the ALD system is relatively low, the uniformity, conformality and the compactness of the film cross-section achieved from the ALD technique are superior to those from other techniques [20].

In this work, we demonstrate the application of AZO thin films as a substitute for ITO electrodes in LC and PDLC display devices. Using the ALD technique, AZO films are deposited on glass and PET substrates and their structural, optical and electric properties are measured and discussed. AZO layers on PET substrates show good flexing properties, as evidenced by their stable sheet resistance over 1000 bending cycles.

The article is structured as follows: we first report on the ALD deposition of AZO films, followed by brief characterizations of the layers, including structural, optical and electrical properties. The next step includes the measurement of the bending ability of the AZO layer deposited on a flexible PET substrate.

In the final section, on the basis of the above characteristics, we present two applications of AZO as a transparent and conductive layer, respectively, in (i) a liquid crystal (LC) display using AZO/glass and (ii) flexible polymer-dispersed liquid crystal (PDLC) devices using AZO/PET. The obtained results show the great potential of AZO for integration into next-generation ITO-free flexible and stretchable devices.

2. Materials and Methods

Al-doped ZnO films were deposited using an ALD Beneq TFS-200 system (Espoo, Finland). Trimethylaluminum (TMA) and diethylzinc (DEZ) were purchased from Strem Chemicals, Inc. (Bischheim, France), and used as received. These precursors and distilled water were held at room temperature in stainless steel containers for all depositions. Purge and line flows used pure N_2 (600 sccm for TMA, DEZ, and water). The precursors' pulse durations were the same for DEZ, TMA, and H_2O—200 ms, while the purging time after each precursor was 2 s, for all deposition runs. In all depositions, the TMA pulse was introduced after a DEZ pulse/purge cycle, in order to minimize the impact of the TMA pulse on the growth rate [21]. The Al-doping of ZnO was controlled by varying the number of DEZ/H_2O and TMA/H_2O pulses [22]. In a typical deposition procedure, after each 24 cycles of DEZ/H_2O, a cycle of TMA/H_2O was applied consisting one so-called supercycle. The desired film thickness was controlled by the number of supercycles. The depositions were performed with the substrate holder temperature of 200 °C for AZO/glass and 100 °C for AZO/PET. The structural, optical and electrical properties of ALD Al-doped ZnO films were measured by using Atomic Force Microscopy (AFM), Scanning Electron Microscopy (SEM), spectrophotometry and a four-point probe method.

Additionally, in the case of AZO deposition on PET substrates, a buffer film of amorphous Al_2O_3 (ALO) was deposited in situ at the same temperature of 100 °C. The buffer film prevents the diffusion of the next precursors in the pores of the PET substrate. For ALO buffer deposition, the same precursors (TMA/H_2O) were used with the same pulse durations and purging times and 88 repetitions of the TMA/H_2O cycle.

The AZO film thickness was determined by ellipsometric measurements using a Woollam M2000D rotating compensator spectroscopic ellipsometer (Lincoln, MI, USA) with a wavelength range from 193 to 1000 nm in reflection mode. For AZO/glass, the thickness was 206 nm and for AZO/PET the thickness was approximately 100 nm. The Al_2O_3 buffer film deposited on PET substrates as barrier to prevent diethylzinc diffusion into the polymer substrate [23] was approximately 15-nm thick.

The AZO films' surface morphology was analyzed using Atomic Force Microscopy (AFM), MFP-3D, Asylum Research, Oxford Instruments (Abingdon, UK) and Scanning Electron Microscopy (SEM), JEOL (Tokyo, Japan).

The optical transmittance spectra of AZO films deposited on glass and PET in the wavelength range of 200 to 1600 nm were measured at room temperature using an ultraviolet–visible–near-infrared (UV–VIS–NIR) spectrophotometer Cary 5E (Palo Alto, CA, USA). The same measurements were performed on reference glass, PET and ITO/glass substrates for comparison.

The sheet resistance was measured using a four-point probe technique. A computerized home-built bending setup, with an ESP301 control platform (Newport, Irvine, CA, USA) with the function to stimulate bending at different radii, was used for the bending tests of flexible samples. The sheet resistance was tested in the interval of up to 1000 bending cycles.

Several AZO films deposited on glass and PET were selected for the LC and PDLC device assembly. The LC and PDLC device fabrication procedure is described in detail in Section 3.2: Applications. The electro-optical characteristics of fabricated LC and PDLC devices were measured by positioning the assembled cell in the optical setups, also described in detail in the same section.

3. Results and Discussion

For an application as a transparent conductive layer, the high optical transmission and high electrical conductivity of the AZO layers are the most essential properties. These properties also depend on the film's elemental composition (doping concentration), deposition conditions/method and the interface with the substrate.

3.1. AZO Layers Characterisations

3.1.1. Structural Analysis

The surface morphologies of the samples were analyzed by using the AFM method with a scanning area of 40 × 40 μm^2, as well as SEM. AFM and SEM images for the surface morphologies of AZO films on glass and PET (before and after bending) are shown in Figure 1a–f. The surface of the films on glass was reasonably smooth, with a roughness of < 3 nm. Grain growth was dense with highly homogenous distribution, as clearly seen in the 3-D AFM image of the surface. It is well known that the grain size, the distribution and the surface roughness are influenced by the sample thickness [16]. The measured root mean square (RMS) roughness of AZO/glass was about 2.3 nm. As seen from the SEM, the as-grown AZO films on both substrates–glass and PET are relatively smooth, with a uniform distribution of slightly elongated grains. The average grain size of AZO films is approximately the same, around 20 nm. The RMS roughness of AZO/PET and SEM morphology before and after bending are discussed in Section 3.1.4.

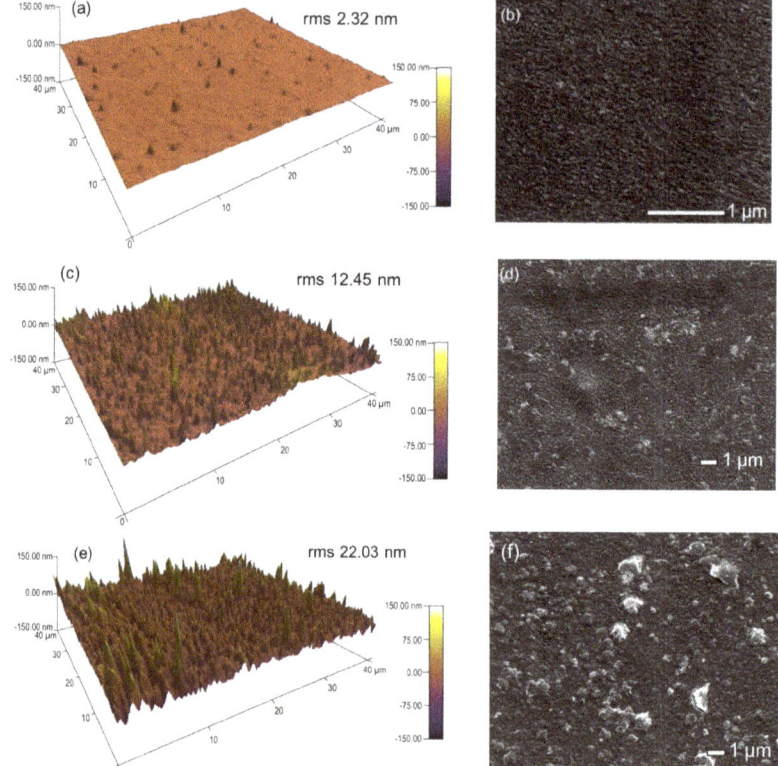

Figure 1. Atomic Force Microscopy (AFM) and SEM analysis of aluminum-doped zinc oxide (AZO) layers deposited on (**a**,**b**) AZO/glass (**c**,**d**) AZO/polyethylene terephthalate (PET) (before bending test) and (**e**,**f**) AZO/PET (after bending 1000 times, discussed in 3.1.4.).

3.1.2. Optical Properties

The optical transmittance spectra of AZO films deposited on glass and PET are shown in Figure 2. The spectra of glass and PET blank substrates are also included for reference as well as commercially available ITO/glass substrate. As seen in comparison to the blank PET substrate, the absorption edges of AZO films on glass and PET are shifted to the longer wavelengths and follow a similar transmittance behavior (around 80% transmittance in visible and near-infrared regions).

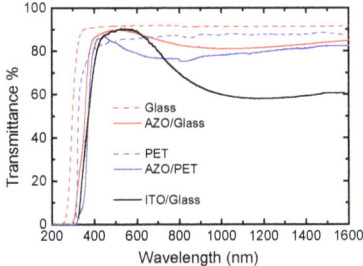

Figure 2. Transmittance spectra of AZO/glass and AZO/PET. Blank glass and PET spectra are also shown for reference, as well as commercial indium tin oxide (ITO)/glass.

3.1.3. Electrical Properties

Next, the electrical properties were assessed. Room temperature currents versus voltage (I–V) characteristics of AZO thin films on glass and PET as well as ITO/glass are shown in Figure 3a–c. All performed current–voltage measurements show Ohmic behavior. As a device-relevant parameter, sheet resistance (R_s) = $1/\sigma d$, where σ is the electrical conductivity of the material and d is the thickness of the electrode, is used for a comparison of the transparent electrodes rather than conductivity, which is an intrinsic material property. The calculated sheet resistances are also shown in the figures.

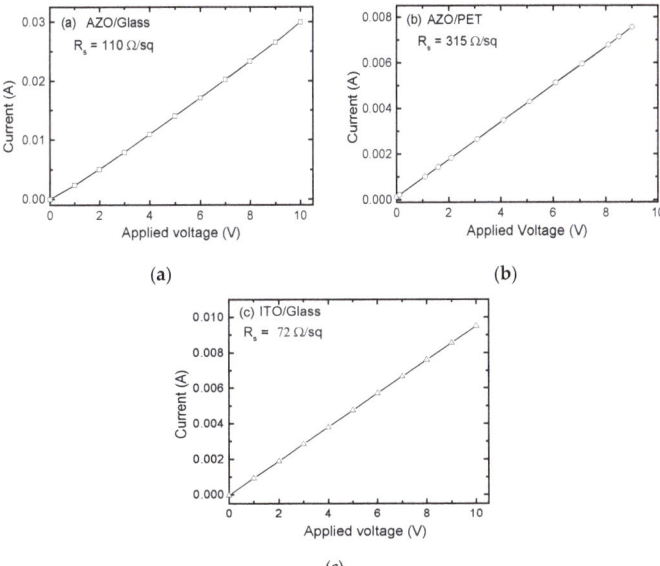

Figure 3. Current–voltage characteristics of AZO thin films deposited on (**a**) glass and (**b**) PET, as well a comparison with (**c**) commercial ITO/glass, are shown for reference.

3.1.4. Bending Ability Test of AZO/PET Flexible Substrates

The bending ability test has been checked for flexible AZO/PET substrates under tensile strain. Generally, the sheet resistance (R_s) value depends on the film composition, morphology, thickness and the substrate material. After continuous bending up to 1000 times (radius of 10 mm), the sheet resistance of AZO/PET increased only slightly (see Figure 4). The RMS roughness of blank PET substrate was 6.26 nm. AZO/PET RMS was measured to be around 12 nm and it is increasing to 22 nm after 1000 bending cycles. During the bending cycles, damage in the AZO/PET develops, most probably in the form of extended defects that collapse as micro-cracks (as evidenced by SEM image in Figure 1f), capable of interrupting electrical paths on subsequent bending. Bending in–out cycles open and close these micro-cracks, and with the accumulation of the bending cycles an effective breaking of a slightly increasing number of conductive paths occurred, thus influencing the measured resistance and appearing as increased sheet resistance. These results are confirmed further by the increased roughness shown in the AFM image of the same layer, performed after 1000 bending cycles (Figure 1c).

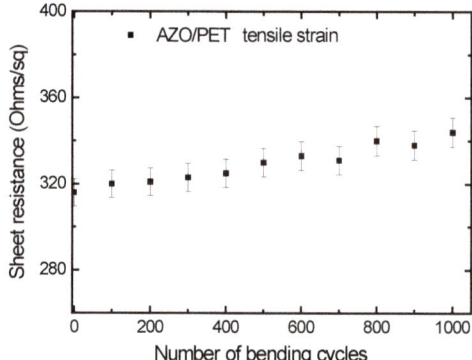

Figure 4. Sheet resistance dependence of the number of bending cycles for AZO layer on PET substrate.

3.2. Applications of AZO Transparent Conductive Layer for Display Applications

3.2.1. AZO as Transparent Conductive Layer in Rigid LC Display

Several AZO/glass samples were selected for transparent electrodes in the LC cell assembly, hereafter called an AZO-based LC rigid display. First, a Polyvinyl Alcohol (PVA) layer was spin-coated, baked at 90 °C for 30 min and post baked at 120 °C for 30 min on AZO/glass samples. Next, the PVA-coated AZO/glass substrates were mechanically rubbed for LC molecule alignment. We used nematic LC (NLC type E7, Merck, Kenilworth, NJ, USA). The cell was arranged by gluing the two rubbed layers in an anti-parallel configuration with the alignment directions facing each other. A 12-μm Mylar spacer was used to keep the same thicknesses in the fabricated cells. The liquid crystal was injected via the capillary method into the empty cell. Finally, the arranged cell was sealed with ultraviolet (UV) glue and exposed with UV light to stabilize it. Copper tape wires were used for electrical connections (see Figure 5a,b). In addition, a reference LC cell using ITO/glass was prepared following the same fabrication procedure and thickness for comparison.

For electro-optical modulation characteristic measurements, the LC cells were placed between a pair of crossed polarizers. The incident beam was polarized at the angle of 45° with respect to the nematic director (a dimensionless unit vector n, which represents the direction of the preferred orientation of LC molecules). An alternating voltage (f = 1 kHz) with varying amplitude was applied across to the LC cell to orient the LC director. A helium–neon laser (He–Ne) emitting λ = 633 nm was used to probe the changes in the transmitted intensity of LC devices under different amplitudes of driving voltage. The power of the transmitted beam was monitored by a power meter. The light transmission was measured as a function of the applied root mean square (RMS) alternating current (AC) voltage with a 1-kHz frequency. The transmittance changes were detected by positioning a photodetector behind the device (Figure 5d).

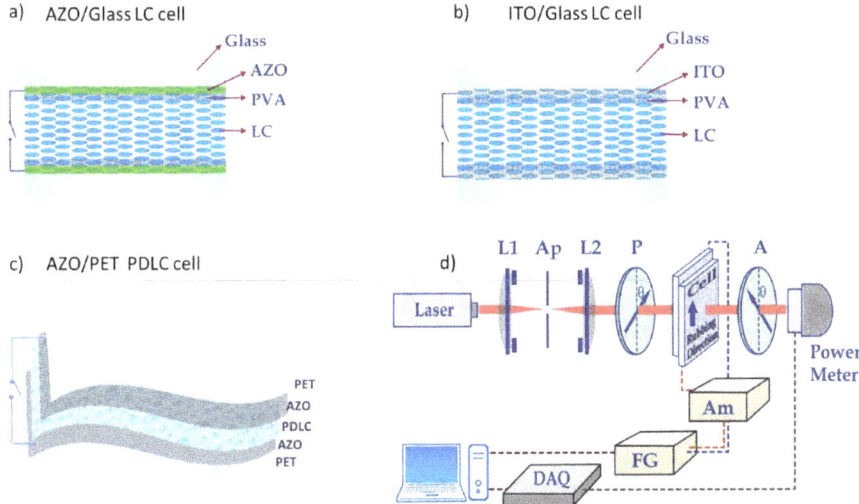

Figure 5. Schematic diagrams of: (**a**) AZO/glass LC cell; (**b**) ITO/glass LC cell and (**c**) AZO/PET polymer-dispersed liquid crystal (PDLC) cell and (**d**) experimental setup to measure the electro-optical modulation behavior. Legend: lens (L), aperture (Ap), polarizer (P), analyzer (A), amplifier (Am), function generator (FG), data acquisition card (DAQ). Note: for PDLC cell measurements, the P and A are removed from the setup.

The basic operation principles of LC displays rely on the electro-optically controlled birefringence of the LC molecules [24,25]. Upon application of an electric field, since the LC molecules have a different polarizability along their long and short axis, an induced dipole moment arises and all the molecules start to reorient towards the direction of the applied field. Because of the LC's anisotropy, the nematic LC layer acts as the birefringent material, characterized by different refractive indices for a beam polarized along the long or short molecular axis.

Figure 6a,b shows the transmitted light intensity dependence of the applied voltage for an LC cell using AZO as transparent electrode. A reference cell, assembled using a commercial ITO electrode is shown for comparison. The transmittance–voltage behavior follows the typical sinusoidal function of the amplitude of applied voltage, governed by the electrical response of the liquid crystal molecules, as expressed by [25].

$$T = \frac{1}{2}\sin^2\frac{\Gamma}{2} = 1/2\sin^2\left[\frac{\pi d(n_e - n_0)}{\lambda}\right] \quad (1)$$

where $\Gamma = \frac{2\pi}{\lambda}(n_e - n_0)d$ is the phase retardation due to the birefringence Δn ($\Delta n = n_e - n_0$) modulation, λ—the wavelength, d—the thickness of the LC layer and n_e and n_0 are the refractive indices for extraordinary and ordinary waves, respectively.

As shown in Figure 6a, the transmitted intensity follows a series of maxima and minima (so-called Fréedericksz transition), which correspond to the phase retardation, as presented in Figure 6b.

The low sheet resistance of the AZO layer has a large impact on the modulation characteristics of assembled LC cells. As seen from Figure 6a, the modulation behavior of the LC cell using AZO contacts is very similar and competitive with the LC cell using ITO. The results are also supported by the calculated Haacke figure of merit (*FOM*) factor of performance, expressed by the ratio between the optical transmittance (T_{av}) and the sheet resistance (R_s) values [26]:

$$\text{FOM} = \frac{T_{av}^{10}}{R_s} \qquad (2)$$

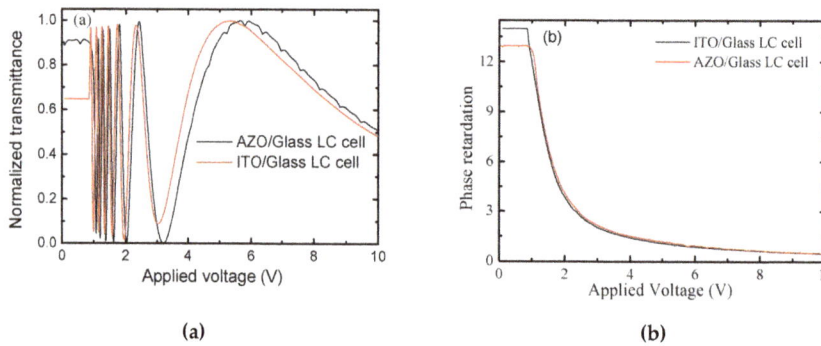

Figure 6. (a) Transmittance–voltage characteristics of AZO/glass and ITO/glass (reference) LC cells and (b) phase retardation for both LC cells.

Using the transmittance data (Figure 2) and measured sheet resistance values of AZO/glass, we obtained FOM $_{(AZO)}$ = 1.05 × 10^{-3} Ω^{-1}. Similar values have been calculated for (commercially available) ITO/glass, FOM $_{(ITO)}$ = 2.52 × 10^{-3} Ω^{-1}. These values are in good confirmation with those previously reported [27,28] and demonstrate the high potential of AZO films as transparent electrodes for ITO-free LCD devices (Table 1).

Table 1. Haacke figures of merit (FOM).

LC Cell Type	T_{av} (%)	R_s (Ω/sq)	FOM (Ω^{-1})
AZO/Glass	80.6	110	1.05 × 10^{-3}
ITO/Glass	84.3	72	2.52 × 10^{-3}

3.2.2. AZO as Transparent Conductive Layer in Flexible Display (Smart Structures)

PDLC-based structures have attracted increasing attention for applications as outdoor displays, switchable privacy glasses, energy saving windows, light shutters, projection displays, and so on [29,30]. Here, the PDLC film was prepared using a polymerization-induced phase separation method, based on the phase separation of the LC (E7, Merck, n_0 = 1.521 and n_e = 1.72) and photo-curable adhesive polymer matrix (NOA65, Norland, Cranbury, NJ, USA, n = 1.524) with a 3:7 weight ratio. During the fabrication of our flexible PDLC structures, supported by AZO/PET substrates, we perform the following steps: first, an empty cell was prepared by gluing two AZO/PET substrates with 12-μm Mylar spacers between them. Then, the LC/monomer mixture was injected into the empty cell and, finally, the cell was exposed with ultraviolet light ((λ = 365 nm) with an intensity of 60 mW/cm^2 for 15 min) to polymerize NOA65. As a result, using the polymerization-induced phase separation method, randomly dispersed liquid crystal droplets (usually in micrometer size) are formed within a transparent polymer matrix. Figure 5c shows the schematic structure of the AZO/PET PDLC flexible device.

3.2.3. Electro-Optical Characteristics and Bending Test Ability

In general, PDLC can be switched between a light-scattering to a transparent state by applying an external electric field. The effect results from a mismatch or match of refractive indices between the LC molecules and the polymer matrix and is due to the LC birefringence and the ability of an applied voltage to re-orient the LC molecules inside the droplets in order to match the LC's refractive index to that of the polymer matrix. The electro-optical characteristics were measured by an optical setup, shown in Figure 5d. In these measurements, the polarizer and analyzer were removed from the setup, since the polymer defines the polarization state of the PDLC film.

The typical transmittance dependence of the assembled AZO/PET PDLC device as a function of the applied voltage is shown in Figure 7a. Without an applied voltage, the LC molecules are randomly oriented in the droplets, which causes light scattering when the light passes through the structure. As a result, the "scattering" state appears. When the voltage is applied, the electric field aligns the LC's nematic director to the direction of the electric field, allowing light to pass through the droplets. As a result, a "transparent" state appears. Thus, the intensity of the transmitted light through the PDLC structure can be controlled by the application of an external voltage.

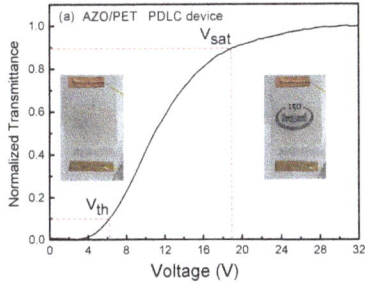
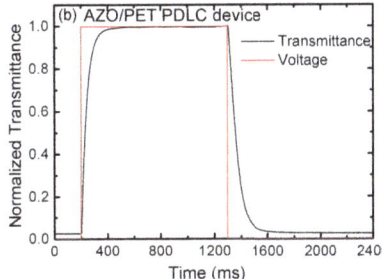

Figure 7. (**a**) Voltage-transmittance behavior of AZO/PET PDLC device and (**b**) the response time.

For the AZO/PET device shown in Figure 7a, we defined (i) the threshold voltage V_{th} as a value of the applied voltage necessary to reach 10% of the maximum transmittance, e.g., to "turn on" the PDLC cell (measured V_{th} ~ 6.1 V) and (ii) the saturation voltage V_{sat}, defined as the value of applied voltage required to reach 90% of maximum transmittance T (V_{sat} ~18 V). As seen, at the saturation state, the PDLC structure becomes transparent (the "BAS LOGO" image pattern become clear).

In addition, the response time of the assembled PDLC device for switching between "off" and "on" states was measured and is presented in Figure 7b. The measured response time and the fall time values were ~68 and ~88 ms, respectively. These values are very similar to the reported values for PDLC devices using other transparent contacts [31,32].

To sum up, the implementation of AZO layers as transparent conductive electrodes in LC and PDLC devices requires the synthesis of high uniformity, conformal and compact layers by an appropriate process such as ALD. The AZO films obtained by ALD technique enable low sheet resistance and high optical transparency through the use as transparent electrodes on the top surfaces of both the selected rigid and flexible substrates.

4. Conclusions

In summary, we studied the structural and optical properties and the sheet resistance ability of AZO layers deposited by the ALD technique onto glass and PET substrates. LC devices with transparent AZO electrodes were successfully fabricated, and the measured driving voltage and response time values show the great potential of AZO for integration as functional transparent conducting material. We expect that the above demonstrations will be relevant for further applications in high-end optoelectronics.

Author Contributions: Conceptualization, V.M., S.H.L., and K.Y.H.; methodology D.D., B.B., S.P., and D.P.; software, S.P., C.-L.T., V.S., and B.N.; formal analysis, V.M., S.H.L., and K.Y.H.; resources D.D., V.M., and S.H.L.; writing—original draft preparation, D.D., and V.M.; writing—review and editing D.D., C.-L.T., V.M., B.B., S.P.,

D.P., V.S., B.N., S.H.L., and K.Y.H.; visualization, D.D., C.-L.T., V.M., S.P., D.P. and V.S.; supervision, V.M., S.H.L., and K.Y.H.; funding acquisition D.D., V.M., and S.H.L. All authors have read and agreed to the published version of the manuscript.

Funding: This research was funded by the Bulgarian Science Fund under the Grant No. KΠ-06-H-28/8 and the Ministry of Science and Technology (MOST), Taiwan, under the Grant No.: MOST 107-2221-E-009-120-MY3 and MOST 109-2927-I-009-507. We are grateful for the support from the Higher Education Sprout Project of the National Chiao Tung University and Ministry of Education (MOE), Hsinchu, Taiwan. Research equipment from the distributed research infrastructure INFRAMAT (part of the Bulgarian National roadmap for research infrastructures), supported by the Bulgarian Ministry of Education and Science under contract D01-284/17.12.2019, was used for AFM measurements.

Acknowledgments: The authors would like to thank P. Terziyska from the ISSP, BAS, Bulgaria for determining the AZO layer's thickness and to Marina Vasileva from the IOMT, BAS, Bulgaria for the measurement of the optical spectra. The authors gratefully acknowledge the financial support provided by the Bulgarian Science Fund under the grant number KΠ-06-H-28/8 and the Ministry of Science and Technology (MOST), Taiwan, under the grant numbers MOST 107-2221-E-009-120-MY3 and MOST 109-2927-I-009-507 (Bulgarian–Taiwanese PPP exchange program). We are grateful for the support from the Higher Education Sprout Project of the National Chiao Tung University and Ministry of Education (MOE), Taiwan. Research equipment from the distributed research infrastructure INFRAMAT (part of the Bulgarian National roadmap for research infrastructures), supported by the Bulgarian Ministry of Education and Science under contract D01-284/17.12.2019, was used for AFM measurements.

Conflicts of Interest: The authors declare no conflict of interest.

References

1. Ellmer, K. Past achievements and future challenges in the development of optically transparent electrodes. *Nat. Photonics* **2012**, *6*, 809–817. [CrossRef]
2. Gordon, R.G. Criteria for choosing transparent conductors. *MRS Bull.* **2000**, *25*, 52–57. [CrossRef]
3. Morales-Masis, M.; De Wolf, S.; Woods-Robinson, R.; Ager, J.W.; Ballif, C. Transparent electrodes for efficient optoelectronics. *Adv. Electron. Mater.* **2016**, *3*, 1600529. [CrossRef]
4. Levy, D.; Castellon, E. *Transparent Conductive Materials: Materials, Synthesis, Characterization, Applications*, 1st ed.; Wiley-VCH Verlag Gmbh &Co, KGaA: Weinheim, Germany, 2018.
5. Transparent Conductive Films Market by Application (Smartphones, Tablets, Notebooks, LCDs, Wearable Devices), Material (ITO on Glass, ITO on PET, Metal Mesh, Silver Nanowires, Carbon Nanotubes), and Region-Global Forecast to 2026. Available online: https://www.marketsandmarkets.com/Market-Reports/transparent-conductive-film-market-59909084.html (accessed on 30 April 2020).
6. Suzuki, A.; Matsushita, T.; Wada, N.; Sakamoto, Y.; Okuda, M. Transparent conducting Al-doped ZnO thin films prepared by pulsed laser deposition. *Jpn. J. Appl. Phys.* **1996**, *35*, L56–L59. [CrossRef]
7. Berginski, M.; Hupkes, J.; Schulte, M.; Schope, G.; Stiebig, H.; Rech, B.; Wuttig, M. The effect of front ZnO:Al surface texture and optical transparency on efficient light trapping in silicon thin-film solar cells. *J. Appl. Phys.* **2007**, *101*, 074903. [CrossRef]
8. Minami, T.; Sato, H.; Nanto, H.; Takata, S. Heat treatment in hydrogen gas and plasma for transparent conducting oxide films such as ZnO, SnO_2 and indium tin oxide. *Thin Solid Films* **1989**, *176*, 277–282. [CrossRef]
9. Dhere, R.G.; Bonnet-Eymard, M.; Charlet, E.; Peter, E.; Duenow, J.N.; Li, J.V.; Kuciauskas, D.; Gessert, T.A. CdTe solar cell with industrial Al:ZnO on soda-lime glass. *Thin Solid Films* **2011**, *519*, 7142–7145. [CrossRef]
10. Vasekar, P.S.; Dhere, N.G.; Moutinho, H. Development of $CIGS_2$ solar cells with lower absorber thickness. *Sol. Energy* **2009**, *83*, 1566–1570. [CrossRef]
11. Park, S.; Ikegami, T.; Ebihara, K. Growth of transparent conductive Al-doped ZnO thin films and device applications. *Jpn. J. Appl. Phys.* **2006**, *45*, 8453–8456. [CrossRef]
12. Verma, A.; Khan, F.; Kumar, D.; Kar, M.; Chakravarty, B.C.; Singh, S.N.; Husain, M. Sol–gel derived aluminum doped zinc oxide for application as anti-reflection coating in terrestrial silicon solar cells. *Thin Solid Films* **2010**, *518*, 2649–2653. [CrossRef]
13. Mondragon-Suarez, H.; Maldonado, A.; de la L Olvera, M.; Reyes, A.; Castanedo-Perez, R.; Torres-Delgado, G.; Asomoza, R. ZnO:Al thin films obtained by chemical spray: Effect of the Al concentration. *Appl. Surf. Sci.* **2002**, *193*, 52–59. [CrossRef]

14. Ma, J.; Ji, F.; Ma, H.; Li, S. Preparation and properties of transparent conducting zinc oxide and aluminium-doped zinc oxide films prepared by evaporating method. *Sol. Energy Mater. Sol. Cells* **2000**, *60*, 341–348. [CrossRef]
15. Liu, Y.; Lian, J. Optical and electrical properties of aluminum doped ZnO thin films grown by pulsed laser deposition. *Appl. Surf. Sci.* **2007**, *253*, 3727–3730. [CrossRef]
16. Yang, W.; Wu, Z.; Liu, Z.; Pang, A.; Tu, Y.; Feng, Z.C. Room temperature deposition of Al-doped ZnO films on quartz substrates by radio-frequency magnetron sputtering and effects of thermal annealing. *Thin Solid Films* **2010**, *519*, 31–36. [CrossRef]
17. Kon, M.; Song, P.K.; Shigesato, Y.; Frach, P.; Mizukami, A.; Suzuki, K. Al-doped ZnO films deposited by reactive magnetron sputtering in mid-frequency mode with dual cathodes. *Jpn. J. Appl. Phys.* **2002**, *41*, 814–819. [CrossRef]
18. Tynell, T.; Yamauchi, H.; Karppinen, M.; Okazaki, R.; Terasaki, I. Atomic layer deposition of Al-doped ZnO thin films. *J. Vac. Sci. Technol. A* **2013**, *31*, 01A109. [CrossRef]
19. Johnson, R.W.; Hultqvist, A.; Bent, S.F. A brief review of atomic layer deposition: From fundamentals to applications. *Mater. Today* **2014**, *17*, 236–246. [CrossRef]
20. Dhakal, T.; Nandur, A.S.; Christian, R.; Vasekar, P.; Desu, S.; Westgate, C.; Koukis, D.I.; Arenas, D.J.; Tanner, D.B. Transmittance from visible to mid infra-red in AZO films grown by atomic layer deposition system. *Sol. Energy* **2012**, *86*, 1306–1312. [CrossRef]
21. Tulzo, H.L.; Schneider, N.; Lincot, D.; Patriarche, G.; Donsanti, F. Impact of the sequence of precursor introduction on the growth and properties of atomic layer deposited Al-doped ZnO films. *J. Vac. Sci. Technol. A* **2018**, *36*, 041502. [CrossRef]
22. Blagoev, B.S.; Dimitrov, D.Z.; Mehandzhiev, V.B.; Kovacheva, D.; Terziyska, P.; Pavlic, J.; Lovchinov, K.; Mateev, E.; Leclercq, J.; Sveshtarov, P. Electron transport in Al-doped ZnO nanolayers obtained by atomic layer deposition. *J. Phys. Conf. Ser.* **2016**, *700*, 012040. [CrossRef]
23. Sweet, W.J., III; Jur, J.S.; Parsons, G.N. Bi-layer Al_2O_3/ZnO atomic layer deposition for controllable conductive coatings on polypropylene nonwoven fiber mats. *J. Appl. Phys.* **2013**, *113*, 194303. [CrossRef]
24. Yeh, P.; Gu, C. *Optics of Liquid Crystal Display*; Wiley Interscience: New York, NY, USA, 2010.
25. Khoo, I.C. *Liquid Crystals Physical Properties and Nonlinear Optical Phenomena*; Wiley: Hoboken, NJ, USA, 1995.
26. Haacke, G. New figure of merit for transparent conductors. *J. Appl. Phys.* **1976**, *47*, 4086–4089. [CrossRef]
27. Oh, B.-Y.; Jeong, M.-C.; Moon, T.-H.; Lee, W.; Myoung, J.-M.; Hwang, J.-Y.; Seo, D.-S. Transparent conductive Al-doped ZnO films for liquid crystal displays. *J. Appl. Phys.* **2006**, *99*, 124505. [CrossRef]
28. Su, Y.C.; Chiou, C.C.; Marinova, V.; Lin, S.H.; Bozhinov, N.; Blagoev, B.; Babeva, T.; Hsu, K.Y.; Dimitrov, D.Z. Atomic layer deposition prepared Al-doped ZnO for liquid crystal displays applications. *Opt. Quantum Electron.* **2018**, *50*, 205. [CrossRef]
29. Baetens, R.; Jelle, B.P.; Gustavsen, A. Properties, requirements and possibilities of smart windows for dynamic daylight and solar energy control in buildings: State-of-the-art. *Sol. Energy Mater. Sol. Cells* **2010**, *94*, 8–105. [CrossRef]
30. Chung, S.H.; Noh, H.Y. Polymer-dispersed liquid crystal devices with graphene electrodes. *Opt. Express* **2015**, *23*, 32149. [CrossRef]
31. Sannicolo, T.; Lagrange, M.; Cabos, A.; Celle, C.; Simonato, J.P.; Bellet, D. Metallic Nanowire-based transparent electrodes for next generation flexible devices: A review. *Small* **2016**, *12*, 6052–6075. [CrossRef]
32. Huang, Q.; Shen, W.; Fang, X.Z.; Chen, G.; Yang, Y.; Huang, J.; Tan, R.; Song, W. Highly thermostable, flexible, transparent, and conductive films on polyimide substrate with an AZO/AgNW/AZO structure. *ACS Appl. Mater. Interfaces* **2015**, *7*, 4299–4305. [CrossRef]

© 2020 by the authors. Licensee MDPI, Basel, Switzerland. This article is an open access article distributed under the terms and conditions of the Creative Commons Attribution (CC BY) license (http://creativecommons.org/licenses/by/4.0/).

Article

Low-Temperature Fabrication of High-Performance and Stable GZO/Ag/GZO Multilayer Structures for Transparent Electrode Applications

Akmedov K. Akhmedov [1], Aslan Kh. Abduev [1], Vladimir M. Kanevsky [2], Arsen E. Muslimov [2] and Abil Sh. Asvarov [1,2,*]

[1] Institute of Physics, Dagestan Federal Research Center, Russian Academy of Sciences, 367015 Makhachkala, Russia; cht-if-ran@mail.ru (A.K.A.); a_abduev@mail.ru (A.Kh.A.)
[2] Shubnikov Institute of Crystallography, Federal Scientific Research Center Crystallography and Photonics, Russian Academy of Sciences, 119333 Moscow, Russia; Kanevsky_V_M@mail.ru (V.M.K.); amuslimov@mail.ru (A.E.M.)
* Correspondence: abil-as@list.ru

Received: 13 February 2020; Accepted: 12 March 2020; Published: 13 March 2020

Abstract: Presently, research and development of indium-free stable highly transparent conductive (TC) materials is of paramount importance for the blooming world of information display systems and solar energy conversion. Development of devices based on flexible organic substrates further narrows the choice of possible TC materials due to the need for lower deposition and process temperatures. In our work, the structural, electrical, and optical performances of Ga-doped ZnO/Ag/Ga-doped ZnO (GZO/Ag/GZO) multilayered structures deposited on glass substrates by direct current (DC) magnetron sputtering in a pure Ar medium without any purposeful substrate heating have been investigated. The highest figure of merit achieved was $5.15 \times 10^{-2}\ \Omega^{-1}$ for the symmetric GZO/Ag/GZO multilayer, featuring GZO and Ag thicknesses of 40 and 10 nm, respectively, while the average optical transmittance was over 81% in the visible range of wavelengths and the resistivity was $2.2 \times 10^{-5}\ \Omega \cdot cm$. Additionally, the good durability of the performances of the multilayer structures was demonstrated by their testing in the context of long-term storage (over 500 days) in standard environmental conditions.

Keywords: multilayer; ZnO; Ag; TCO; transmittance; structure; resistance

1. Introduction

Transparent electrodes (i.e., thin films based on transparent conductive (TC) materials) are some of the most important parts of many optoelectronic devices, such as touch panels, organic light-emitting diodes (OLEDs), optical sensors, and solar cells [1–5].

Nowadays, transparent electrodes based on Sn-doped In_2O_3 (ITO) present outstanding optoelectronic performance and have been widely used in various commercial domains [1,6]. However, the wide use of ITO transparent electrodes in optoelectronic devices is gradually pushing up the cost of ITO electrodes because indium is not abundant on Earth. Moreover, with the rapid development of new types of display systems, sensors, and solar energy, new requirements for transparent electrodes are emerging from device developers, in addition to their transparency and conductivity. It is getting harder for the traditional ITO electrodes to meet the new requirements. Therefore, alternative materials should be developed.

A variety of ITO replacements have been investigated, including doped wide-bandgap oxides with high transmittance, such as SnO_2 [7], ZnO [8,9], and TiO_2 [10]. However, these oxides were found to have lesser performance than ITO, combining both electrical and optical properties. As alternatives to ITO, poly(3,4-ethylenedioxythiophene):poly(4-styrenesulfonate) (PEDOT–PSS) [11], graphene [12], carbon nanotubes [13], and metal nanowires and meshes [14,15] have been proposed. However, each alternative solution is affected by one or more drawbacks that prevent their widespread use [16,17].

In order to keep a low resistance and conversely maintain high optical transmittance, oxide–metal–oxide multilayered structures have recently received renewed interest as a highly promising route towards the production of flexible large area OLEDs and solar cells [3,18–20]. In this case, Ag is the optimal metal because of its low resistivity (approximately 1.6×10^{-6} Ω·cm) and relatively low cost [21], whereas Ga-doped ZnO (GZO) is the optimal oxide due to its abundance, low cost, superior optical features, and rather high stability [22,23].

Various deposition techniques have been used to produce oxide–metal–oxide structures, including thermal evaporation [24], electron beam evaporation [25], spray pyrolysis [26], sol–gel methods [27], ion beam sputtering [28], and magnetron sputtering [18–20,28]. Low-cost wet chemical methods are usually the starting point and benchmark for most academic and industrial processes that require a thin and uniform coating, but the transparent electrodes obtained by these techniques have resulted in inferior electrical features compared with those deposited by ion plasma methods [25,29,30]. In this sense, it appears that DC magnetron sputtering is the most promising technique in terms of the industrial deposition of uniform films at a proper deposition rate [18]. From the point of view of the deposition of transparent electrodes on flexible substrates covering a large area, it is also very important to achieve TC films with good performance stability by using low-temperature processes [28,31].

In this article, symmetric GZO/Ag/GZO (GAG) multilayered structures were sequentially formed on glass substrates by room temperature DC magnetron sputtering under a pure Ar atmosphere. The uniqueness and novelty of this work resides in having found the process conditions that provide the optimal trade-off between low resistivity and high optical transmittance and are applicable for TC thin film formation on polymer substrates. The thicknesses of GZO and Ag layers were parametrized to get the optimal optical and electrical properties of the superstructures. The deposited multilayers were characterized and tested for their structural, electrical, optical, and adhesive properties.

2. Materials and Methods

GAG multilayered structures were deposited on glass and surface-oxidized Si pieces by DC magnetron sputtering method using a Magnetron setup (Russia) equipped with two sputtering units. The base pressure of the chamber was maintained at 2×10^{-4} Pa.

The bottom and top oxides layers were deposited using Ga (3 at.%)-doped ZnO target under the following deposition conditions: Ar working pressure of 0.5 Pa, discharge current of 270 mA, discharge voltage of 720 V. The Ag thin interlayer was deposited using an Ag (99.99%) pure target under the following conditions: Ar working pressure of 0.5 Pa, discharge current of 150 mA, discharge voltage of 750 V. Before the formation of each layer of the GAG trilayered structure, a presputtering cycle of both GZO and Ag targets on a closed shutter was performed for 10 and 3 min, respectively. The distance between targets and substrates was 150 mm. The substrates revolved at a rate of 30 r/min. The substrate was not specially heated during the growth of multilayers, but there was a slight heating to about 50 °C due to ion bombardment. The growth rates for the oxide and metal components of the three-layer structure under such conditions were 1.33 and 3.00 nm/min, respectively. The thickness of oxide and metal components of the GAG multilayer was controlled by varying the deposition time, which was the same for bottom and top GZO films in order to create a symmetric trilayer structure. Layer thickness variations were also confirmed by cross-sectional scanning electron microscopy (SEM) and were consistent with the estimated depositing times (Figure S1 of Supplementary Materials (SM)).

Table 1 shows thickness data of seven multilayered samples with various geometries. For the first five samples (from GAG-0 to GAG-4), the thicknesses of the silver interlayer were varied between 0 and 12 nm, while keeping fixed thicknesses for oxide top and bottom GZO layers (40 nm). For the samples GAG-5 and GAG-6, the thickness of oxide films was varied, keeping the thickness of the Ag interlayers fixed (10 nm). For comparison, a two-layered GA with a 80-nm thick bottom oxide and thin upper metal layer (10 nm) was deposited additionally.

Table 1. Nomenclature of Ga-doped ZnO/Ag/Ga-doped (GZO/Ag/GZO) multilayered structures.

Sample Name	Thickness of Bottom GZO Layer, nm	Thickness of Ag Interlayer, nm	Thickness of Top GZO Layer, nm
GAG-0	40	0	40
GAG-1	40	6	40
GAG-2	40	8	40
GAG-3	40	10	40
GAG-4	40	12	40
GAG-5	30	10	30
GAG-6	50	10	50
GA	80	10	-

The surface morphologies of the deposited samples were investigated by using a Leo-1450 scanning electron microscope (SEM) (Carl Zeiss) and a Ntegra Prima atomic force microscope (NT-MDT SI). X–ray diffraction (XRD) patterns were collected on a X'PERT PRO MPD diffractometer (PANalytical) with CuKα radiation (λ = 1.5418 Å). The electrical properties were measured by using the four-probe technique (IUS-3, Russia). Optical transmittance spectra were obtained by a UV-3600 optical spectrophotometer (Shimadzu) in the wavelength range of 340–1240 nm. All of the measurements were carried out at room temperature.

3. Results and Discussion

3.1. Surface Morphology and Structural Studies

The surface morphology of the GAG samples was investigated by SEM. Figure 1 shows the typical SEM images of top view surfaces observed in this work. A single layer GAG-0 sample with a total thickness of oxide of 80 nm and a zero thickness Ag interlayer (Figure 1a) consists of well-defined continuous particles of nearly equal lateral size (~ 40 nm) uniformly covering the smooth substrate. After deposition of 10-nm thick Ag layer on the GZO surface, a well-marked change in morphology appears (Figure 1b). Forming a sufficiently continuous layer of silver makes the surface of the structure even smoother, although some nanovoids on the surface of the GA bilayered structure are still present due to the surface performance of the bottom GZO layer. The observed smoothing of the surface indicates that under the above growth conditions, the process of Ag growth should be described in the following scenario:

- At the initial stage of growth, a large number of silver nuclei are formed on the surface of the GZO layer due to limited migration of adatoms;
- The high density of the nuclei contributes to their earlier coalescence into a continuous metal film, covering the GZO surface [32].

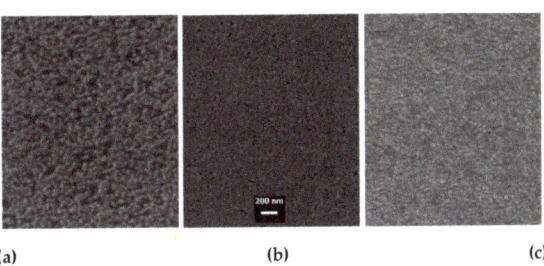

Figure 1. SEM images GAG-0 (a), GA (b), and GAG-3 samples (c).

In earlier reports [33,34], it was found that the spreading of Ag on the surface of ZnO was better than on SnO_2 and In_2O_3 due to an ameliorated affinity between Ag and ZnO. Nevertheless, it was confirmed that the surface of ZnO was rougher than that of the latter.

As is shown in Figure 1c, the surface morphology of GAG-3 trilayered structure became less smooth but quite compact. The difference between the surfaces of GAG-1 and GAG-3 samples is due to both differences in the nucleation conditions of the top and bottom oxide layers, and the fact that the grain size usually increases with the thickness of the ZnO thin film [35,36].

It should be noted that the SEM studies of other samples did not reveal any noticeable differences in the surface morphology of the samples considering the thickness of the Ag interlayer at a fixed thickness of the GZO, as well as when changing the thickness of the GZO layers in the range of 30–50 nm. Thus, we can reach a conclusion that in our SEM experiment, the top GZO, which was sputtered at room temperature, showed surface features typical of nanocrystalline Ga-doped ZnO thin films, regardless of the Ag interlayer thickness and its morphology. At the same time, additional atomic-force microscopy (AFM) studies (Figure S2 in SM) showed that the surface roughness increases noticeably by introducing an Ag interlayer into GZO. The root mean square (RMS) roughness values of GZO-0 and GZO-1 samples are 0.897 and 1.226 nm, respectively, calculated from the AFM data. A slight decrease in roughness is observed with further increases of both Ag and GZO thicknesses in the trilayer structures.

Figure 2a shows the XRD plots of the GAG multilayered structures with Ag interlayers of different thicknesses. Only four broad peaks were present in the XRD spectra, two of which belong to the (002) ZnO and (004) ZnO reflections, and the other two to the (111) Ag and (222) Ag reflections. The presence of two (002) ZnO and (004) ZnO peaks corresponding to the nanocrystalline hexagonal ZnO wurtzite phase indicates that the GZO has a preferential orientation featuring the c-axis perpendicular to the substrate surface, regardless of the Ag interlayer thickness. The insertion of the Ag interlayer in the middle of the GAG structure does not affect the strongly preferred orientation of GZO toward (001). Additionally, for the GAG structures, the Ag interlayer has highly preferred orientation toward (111). It is often reported that the crystallized ZnO lattice promotes the silver growth along the (111) direction. This might be due to the fact that the (111) plane of a cubic structure has a similar symmetry to that of the (001) plane of ZnO [34]. The inset of Figure 2a shows the XRD spectral region in which the most intense (002) ZnO and (111) Ag peaks are located. The main features of both peaks are given in Table 2.

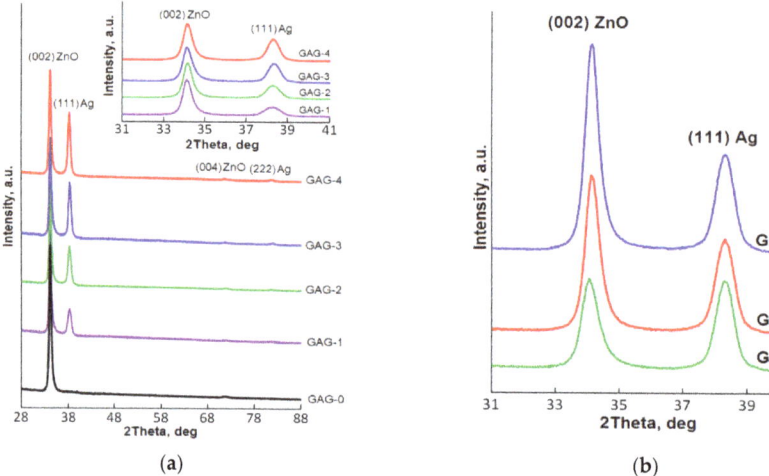

Figure 2. X-ray diffraction (XRD) plots of the prepared multilayer structures: (**a**) with different Ag interlayer thicknesses; (**b**) with different thicknesses for bottom and top GZO layers. The inset of Figure 2a shows the XRD spectral region with the most intense (002) ZnO and (111) Ag peaks.

Table 2. XRD data for GAG multilayer structures.

Sample Name	Thickness of Layers, nm			XRD Data for the (002) ZnO Peak			XRD Data for the (111) Ag Peak			I_{Ag}/I_{ZnO}
	GZO	Ag	GZO	I, cps	β, deg	CS, nm	I, cps	β, deg	CS, nm	
GAG-0	40	0	40	15755	0.520	16	-	-	-	0
GAG-1	40	6	40	9606	0.592	14	2767	1.056	7	0.29
GAG-2	40	8	40	9760	0.602	14	3882	0.964	8	0.40
GAG-3	40	10	40	9728	0.598	14	6045	0.847	10	0.62
GAG-4	40	12	40	9894	0.601	14	6400	0.840	10	0.65
GAG-5	30	10	30	5437	0.665	12	6153	0.829	10	1.13
GAG-6	50	10	50	13240	0.585	15	6750	0.836	10	0.51

It can be seen that when the Ag interlayer is introduced into the GAG structure, thereby breaking the GZO layer into two equal parts by thickness, a decrease in intensity and some broadening of the (002) ZnO peak takes place. A further increase of the thickness of the Ag interlayer until 12 nm does not affect the crystallinity of the GZO phase, which is correlated with the SEM data results.

At the same time, the intensity of the (111) Ag peak increases and the integral breadth β decreases with a thickening of the Ag interlayer. The peak shifts from 38.21 to 38.25°, with an increase in the Ag thickness from 6 to 8 nm, after which its position no longer changes.

Estimation of the averaged crystallite size (CS) from (002) ZnO and (111) Ag peak characteristics using the Scherrer equation (CS = $0.9\lambda/(\beta\cos\theta)$, where λ is the wavelength of CuKα x-rays, β is the peak integral breadth with no instrumental contribution, and θ is the peak Bragg angle) showed that the crystallite size of GZO decreases from 16 to 14 nm when the Ar interlayer is introduced, and the Ag crystallite size increases continuously with the increase of the Ag thickness from 7 to 10 nm.

The change in the ratio of the intensities I of the (111) Ag and (002) ZnO peaks (I_{Ag}/I_{ZnO}) is in agreement with the deposition regimes for these GAG structures.

Despite the low substrate temperature during the GZO and Ag sputtering process, the GAG multilayer structures consist of both nanocrystalline GZO and Ag layers (Figure S3 of SM). In addition, we can reach the conclusion that the crystallinity of the top GZO layer is independent of the Ag interlayer.

GZO thickness variation in the range of 30–50 nm at a fixed thickness of the Ag interlayer of 10 nm does not affect the preferential orientation of both GZO and Ag layers. Figure 2b shows the XRD plots of the GAG structures as a function of the top and bottom GZO thickness in the 2θ range of 31–41°. XRD data for (002) ZnO and (111) Ag peaks of GAG-5 and GAG-6 are shown in Table 2 (in order to compare these with GAG-3).

Comparing these samples, the (002) ZnO peak shifts from 34.01 to 34.10° with an increase in the GZO thickness from 30 to 40 nm. A further increase of the GZO thickness does not change the peak position at 34.10. It can also be observed that as the thickness of GZO film increases, the intensity I is enhanced and the integral breadth β decreases for the (002) ZnO diffraction peak, indicating that the increased thickness of top and bottom layers improves the crystallinity of the GZO phase.

As for the (111) Ag peak, its features for this sample group were practically independent of the thickness of the oxide layers, which implies the invariability of the crystallinity of the Ag interlayer with the increase of the GZO thickness. By increasing the GZO thickness, the averaged crystallite size for GZO increases continuously from 12 to 15 nm, while the Ag crystallite size remains unchanged and remains in good agreement with the thickness of the Ag interlayer.

Thus, we can conclude that the crystallinity of the Ag interlayer is insensitive to changes in the thickness of the bottom GZO layer for the low substrate temperature sputtering process. In this case, the presence of this GZO layer itself is important as a seed layer for Ag. Changes in the nature of coalescence of Ag nuclei were observed in the presence of seed layers with a thickness of only a few nanometers [37,38].

3.2. Optical and Electrical Studies

The GAG-0 sample without any silver interlayer showed a high sheet resistance R_S of 2500 Ω/sq and electrical resistivity ρ in the order of 10^{-2} $\Omega\cdot$cm due to the low preparation temperature. The optical transmittance of GAG-0 is 88% in the visible range of wavelengths (400–700 nm), as shown in Figure 3a, which is consistent with early studies [23,26]. After covering this GZO film with the 10-nm thick Ag layer (GA sample), the resistivity decreases to 2.3×10^{-5} $\Omega\cdot$cm. It is obvious that the conductivity of such a two-layer structure is mainly governed by the continuity and homogeneity of the Ag thin layer [39]. However, as can be seen from Figure 3a, the existence of the Ag layer on top of the 80-nm thick GZO layer substantially reduces the optical transmittance in the visible and near infrared (NIR) regions (the average visible transmittance T_{av} is 41.5%) [28].

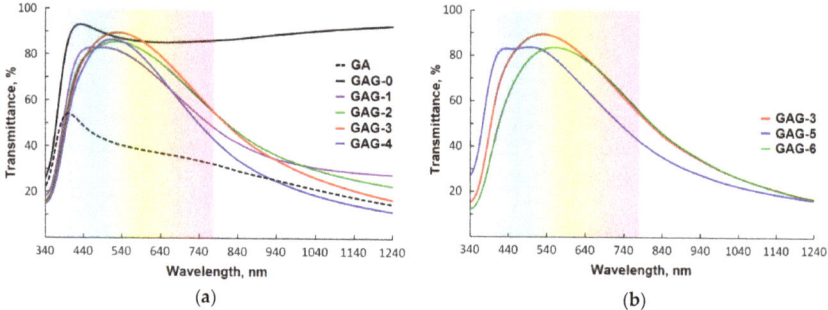

Figure 3. Optical transmittance spectra of the prepared multilayer structures: (**a**) with different Ag interlayer thicknesses; (**b**) with different thicknesses for the bottom and top GZO layers.

In Figure 3a, the optical transmittance spectra are presented of the GAG trilayered structures, consisting of two identical 40-nm GZO layers and Ag interlayers with different thicknesses. As shown in the figure, the top GZO layer antireflects the Ag layer in GAG structures to output higher transmittances than the GA bilayer structure by itself [40]. Moreover, the optical transmittances of the GAG multilayers were found to depend critically on the Ag interlayer thickness.

The average optical transmittance in the visible wavelength T_{av} region of the GAG-1 sample is relatively low (75%) due to light scattering on various defects (pores inherent in ultrathin Ag films and resulting imperfections of the GZO–Ag interfaces). By increasing the Ag thickness to 10 nm, the average optical visible transmittance increases and there is a shift of the transmission peak due to the effects of surface plasmon resonance of the Ag interlayer with minimum voids [41]. In particular, GAG-3 shows high optical transmittance in the visible region, with maximum transmittances of approximately 89% at λ = 529 nm. Next to a wavelength of 550 nm, this sample has the highest optical transmittance, which is even higher than that of GAG-0, which has no silver interlayer.

Further increasing the Ag thickness above 10 nm results in a decrease of the transmittance because of increased light reflection from the continuous Ag interlayer. Additionally, all samples show an abrupt decrease in optical transmittance in the near infra-red region, which is correlated with the thickness of the metal interlayer and attributed to the reflection of long-wavelength light by the layered metal [33]. Thus, the best optical properties of the GAG structures are obtained when the Ag interlayer thickness is 10 nm. The obtained optimal thickness value of the Ag interlayer is similar to the one reported by other groups for ZnO/Ag/Zno multilayers deposited at low substrate temperatures [19,28,42–44].

Figure 3b depicts the transmittance spectra for the GAG structures with the optimum Ag interlayer thickness and different GZO thicknesses. With the increase of GZO thickness, the transmittance first shows an increase and then decreases. Simultaneously, the peak transmittance shifts towards the long wavelength regions. For clarity, the results of our optical measurements are summarized in the corresponding columns of Table 3. Based on these results, the GAG-3 sample with 40-nm thick GZO top and bottom layers and a 10-nm thick Ag interlayer was considered as the optimum choice in terms of optical properties.

Table 3. Optical and electrical parameters of GAG samples.

Sample Name	Thickness of Layers, nm			Optical Data			Electrical Properties		FOM, Ω^{-1}
	GZO	Ag	GZO	T_{av}, %	T_{550nm}, %	λ_{max}, nm	R_S, Ω/sq	ρ, $\times 10^{-5}$ $\Omega \cdot cm$	
GA	80	10	-	41.5	39.9	400	2.2	1.95	6.99×10^{-5}
GAG-0	40	0	40	87.8	86.6	431	2500	2000	1.09×10^{-4}
GAG-1	40	6	40	75.5	80.1	487	15.1	13.0	3.99×10^{-3}
GAG-2	40	8	40	78.8	85.0	526	4.5	4.0	2.04×10^{-2}
GAG-3	40	10	40	81.3	89.0	529	2.45	2.2	5.15×10^{-2}
GAG-4	40	12	40	75.6	84.7	515	2.0	1.85	3.05×10^{-2}
GAG-5	30	10	30	75.3	80.2	498	2.8	2.0	2.08×10^{-2}
GAG-6	50	10	50	74.6	83.6	560	2.2	2.45	2.39×10^{-2}

The results of our study on the dependence of electrical properties of the multilayer samples as sheet resistance resistivity (R_S) and resistivity (ρ) on the thicknesses of the GZO and Ag layers are also presented in Table 3.

According to the presented results, the resistivity ρ can be decreased drastically by three orders of magnitude by inserting a thin Ag interlayer. From the fact that even at the Ag thickness of 6 nm, the specific resistance of the GAG-1 is significantly reduced, it can be assumed that in this case the Ag interlayer is already a continuous network of partially coalesced islands. The material of the top GZO layer partially fills the voids of Ag, therefore additionally shunting the gaps in the metal network. A further increase in the thickness of the Ag interlayer results in both an improvement in the crystalline perfection of the metal phase and a decrease in the size and number of voids in it. Thus, a monotonic decrease in resistance with increasing thickness of the Ag interlayer can be explained by an obvious increase in the total number of charge carriers (the effective carrier concentration) in GAG and also very likely by an increase of carrier mobility.

From the comparison of the electrical properties of the GAG-3, GAG-5, and GAG-6 samples (GZO thickness variation at a fixed Ag thickness of 10 nm), we can verify that the oxide layers in the oxide–Ag–oxide multilayer play only a minor role in the electrical properties of the conductive multilayer structures [45]. While the surface resistance decreases from 2.8 to 2.2 Ω/sq with increasing GZO thickness, there is an increase in resistivity from 2×10^{-5} to 2.45×10^{-5} $\Omega \cdot$cm.

Assuming that the total number of carriers in the metal layer (N_{Ag}) is much greater than the number of carriers in the oxide layer (N_{GZO}), the effective concentration of carriers (n) of the symmetric GAG structure is related to the thickness of the GZO layers by the following expression [46]:

$$n \sim N_{Ag} / (2 \times d_{GZO} + d_{Ag}) \qquad (1)$$

where d_{Ag} and d_{GZO} are the thickness of the metal interlayer and top (bottom) oxide layer, respectively. From this relation, it can be clearly seen that the carrier concentration should be decreased as the GZO layer thickness increases. This is consistent with the above experiment results.

To evaluate the performance of transparent conductive films for various applications, the optical transmission and the electrical conduction of the films should not be considered separately. Simultaneous optimization of low resistivity and transparencies is needed. Usually, the objective evaluation can be carried out using Haacke's figure of merit (FOM) [47], defined as:

$$FOM = T^{10} / R_S \qquad (2)$$

where T is the transmittance at λ = 550 nm or the average visible transmittance. In the last column of Table 3, there are *FOM* values, which are calculated by using the value of the average visible transmittance T_{av} for the all deposited samples. From samples of Ag of varying thickness, the maximum *FOM* of 5.15×10^{-2} Ω^{-1} corresponds to GAG-3, with Ag thickness measuring 10 nm and sheet resistance of 2.45 Ω/sq. This is despite the fact that GAG-4, with Ag thickness of 12 nm, showed a record low resistance (2.0 Ω/sq.). As can be seen, GAG-3 also demonstrates the maximum *FOM* value when comparing samples with the same Ag layer thickness. Additionally, for this sample the *FOM'* was also calculated by using the value T_{550nm} of the transmittance at λ = 550 nm (this parameter is usually used to characterize transparent electrodes for LED and information display applications). The value of *FOM'* is equal 1.27×10^{-1} Ω^{-1} due to the highest transmittance at 550 nm. Both values of the figure of merit of GAG-3 are superior to many of those reported in the literature [19,44–46]. This may be due to the better spreadability of the Ag layer on the GZO layer during DC magnetron deposition at low temperature, where the bottom GZO layer enhances the silver thin film crystallite size [34,48]. This prompts the formation of a uniform and continuous Ag layer at a much thinner thickness, thereby significantly improving its transparency and conductivity characteristics.

3.3. Adherence and Durability Tests

In order to test the adherence of the GAG structures, the scotch tape test was carried out for all the deposited samples. All the GAGs were found to withstand the scotch tape test as soon as prepared and after 500 days of exicator-free storage in paper envelopes. Long-term indoor storage at an average annual humidity and temperature of 65% and 24 °C, respectively, did not lead to any deterioration of the performance of our GAG samples, whereas in the bilayer GA sample, a lot of white dots and spots with fractal-like structures appeared after 45 days [49]. After this, its optical and electrical performance became unacceptable. This result reveals an additional important role of the upper GZO layer as a protective layer, blocking the damaging effects of moisture on the thin Ag layer. Thus, the trilayer GAG structures have good long-term durability and their adherence to the substrate is good.

4. Conclusions

The GZO/Ag/GZO multilayer structures were deposited sequentially by using DC mode only in the magnetron sputtering for both oxide and metal components of the multilayer structure in pure Ar medium and without any purposeful substrate heating. We investigated the structural, electrical, and optical properties of multilayer structures deposited in various combinations of thicknesses of the Ag interlayer and GZO layers. Comparison between XRD, SEM, and electro-optical performance data with each other, as well as with the data from other authors, allows us to draw the following conclusions:

- The bottom nanocrystalline GZO layer contributes to the earlier formation of the continuous Ag layer with highly preferred orientation toward(111);
- The earlier coalescence of Ag nanocrystallites makes it possible to achieve high conductivity for the ultra-thin metal interlayer, characterized by low scattering and lowered plasmonic and intrinsic absorption;
- The top GZO layer, in addition to the antireflection effect, also acts as a protective layer, effectively blocking the interaction of the nano-Ag phase in the presence of external humidity.

The highest FOM value was 5.15×10^{-2} Ω^{-1} for the symmetric GZO/Ag/GZO multilayer with GZO and Ag thicknesses of 40 and 10 nm, respectively, and was achieved when optimizing the geometry of the multilayered structure. This multilayer structure has an average visible transmittance of above 81% and resistivity of 2.2×10^{-5} $\Omega \cdot$cm, values that were unchanged after 500 days storage in a normal environment. In conclusion, using only DC mode in magnetron sputtering and the absence of substrate heating during sample preparation in the context of this work makes our results very promising in terms of further industrial compatibility.

Supplementary Materials: The following are available online at http://www.mdpi.com/2079-6412/10/3/269/s1, Figure S1: Cross-sectional SEM images of the GAG-3 (a) and GAG-4 (b) trilayer structures, Figure S2: AFM images (2×2 µm^2) prepared in semi-contact mode for GAG-0 (a), GAG-1 (b), GAG-3 (c), GAG-4 (d), GAG-5 (e), and GAG-6 (f), Figure S3: GAG-3 XRD plot in comparison with XRD spectra of pure bulk Ag and pure bulk ZnO materials.

Author Contributions: A.K.A. and A.Sh.A. carried out most of the deposition and testing experiments. A.K.A., A.E.M. and A.Sh.A. performed the investigations of samples properties. A.Sh.A., A.Kh.A. and V.M.K. analyzed the data. Writing—original draft preparation, A.K.A. and A.Sh.A. Writing—review and editing, A.Sh.A. In this study, A.Kh.A. and V.M.K. provided the financial and technical support for designing and conducting the research, as well as supervised the whole research process. All authors have read and agreed to the published version of the manuscript.

Funding: This research was performed in the frame of state assignments of Ministry of Science and Higher Education of the Russian Federation for Dagestan Federal Research Center of Russian Academy of Sciences (Dagestan FRC of RAS) and Federal Scientific Research Center "Crystallography and Photonics" of Russian Academy of Sciences (FSRC "Crystallography and Photonics" RAS) and partially funded by Russian Foundation for Basic Research (research project no. 18-29-12099 and no. 19-07-00537). Access to the equipment of the Shared Research Center of FSRC "Crystallography and Photonics" RAS was supported by the Ministry of Science and Higher Education of the Russian Federation (project RFMEFI62119X0035).

Acknowledgments: The authors are grateful for additional technical support from the Shared Research Centers of Dagestan FRC of RAS. The authors acknowledge Alessandro Chiolerio for useful discussions and good recommendations during performing and preparation of the presented work.

Conflicts of Interest: The authors declare no conflict of interest. The funders had no role in the design of the study; in the collection, analyses, or interpretation of data; in the writing of the manuscript, or in the decision to publish the results.

References

1. Sakamoto, K.; Kuwae, H.; Kobayashi, N.; Nobori, A.; Shoji, S.; Mizuno, J. Highly flexible transparent electrodes based on mesh-patterned rigid indium tin oxide. *Sci. Rep.* **2018**, *8*, 2825. [CrossRef] [PubMed]
2. Bi, C.; Chen, B.; Wei, H.; DeLuca, S.; Huang, J. Efficient Flexible Solar Cell based on Composition-Tailored Hybrid Perovskite. *Adv. Mater.* **2017**, *29*, 1605900. [CrossRef] [PubMed]
3. Wu, C.C. Highly flexible touch screen panel fabricated with silver-inserted transparent ITO triple-layer structures. *RSC Adv.* **2018**, *8*, 11862–11870. [CrossRef]
4. Moon, H.; Won, P.; Lee, J.; Ko, S.H. Low-haze, annealing-free, very long Ag nanowire synthesis and its application in a flexible transparent touch panel. *Nanotechnology* **2016**, *27*, 295201. [CrossRef]
5. Rana, A.S.; Lee, J.Y.; Hong, Y.P.; Kim, H.S. Transient Current Response for ZnO Nanorod-Based Doubly Transparent UV Sensor Fabricated on Flexible Substrate. *pss (RRL)* **2018**, *12*, 1800001. [CrossRef]
6. Baraton, M. The Future of TCO Materials: Stakes and Challenges. *MRS Proceedings* **2009**, *1209*, 1209-P03-06. [CrossRef]
7. Banyamin, Z.Y.; Kelly, P.J.; West, G.; Boardman, J. Electrical and Optical Properties of Fluorine Doped Tin Oxide Thin Films Prepared by Magnetron Sputtering. *Coatings* **2014**, *4*, 732–746. [CrossRef]
8. Mickan, M.; Helmersson, U.; Horwat, D. Effect of substrate temperature on the deposition of Al-doped ZnO thin films using high power impulse magnetron sputtering. *Surf. Coat. Tech.* **2018**, *347*, 245–251. [CrossRef]
9. Abduev, A.K.; Akhmedov, A.K.; Asvarov, A.S. UV-assisted growth of transparent conducting layers based on zinc oxide. *Tech. Phys. Lett.* **2017**, *43*, 1016–1019. [CrossRef]
10. Hitosugi, T.; Yamada, N.; Nakao, S.; Hirose, Y.; Hasegawa, T. Properties of TiO2-based transparent conducting oxides. *Phys. Stat. Sol. (a)* **2010**, *207*, 1529–1537. [CrossRef]
11. Lee, J.H.; Jeong, Y.R.; Lee, G.; Jin, S.W.; Lee, Y.H.; Hong, S.Y.; Park, H.; Kim, J.W.; Lee, S.-S.; Ha, J.S. Highly Conductive, Stretchable, and Transparent PEDOT:PSS Electrodes Fabricated with Triblock Copolymer Additives and Acid Treatment. *ACS Appl. Mater. Interfaces* **2018**, *10*, 28027–28035. [CrossRef] [PubMed]
12. Fernández, S.; Boscá, A.; Pedrós, J.; Inés, A.; Fernández, M.; Arnedo, I.; González, J.P.; de la Cruz, M.; Sanz, D.; Molinero, A.; et al. Advanced Graphene-Based Transparent Conductive Electrodes for Photovoltaic Applications. *Micromachines* **2019**, *10*, 402. [CrossRef] [PubMed]
13. Lee, K.-T.; Park, D.H.; Baac, H.W.; Han, S. Graphene- and Carbon-Nanotube-Based Transparent Electrodes for Semitransparent Solar Cells. *Materials* **2018**, *11*, 1503. [CrossRef] [PubMed]
14. Wang, Y.; Du, D.; Yang, X.; Zhang, X.; Zhao, Y. Optoelectronic and Electrothermal Properties of Transparent Conductive Silver Nanowires Films. *Nanomaterials* **2019**, *9*, 904. [CrossRef]

15. Chae, K.S.; Hong, Y.K.; Kim, H.J.; Jeong, J.Y.; Han, T.H. Design of Metal-mesh Electrode-based Touch Panel for Preventing Back-surface Touch Error. *Sens. Mater.* **2019**, *31*, 587–593. [CrossRef]
16. Marciniak, S.; Crispin, X.; Uvdal, K.; Trzcinski, M.; Birgerson, J.; Groenendaal, L.; Louwet, F.; Salaneck, W.R. Light induced damage in poly(3,4-ethylenedioxythiophene) and its derivatives studied by photoelectron spectroscopy. *Synth. Met.* **2004**, *141*, 67–73. [CrossRef]
17. Khaligh, H.H.; Xu, L.; Khosropour, A.; Madeira, A.; Romano, M.; Pradére, C.; Tréguer-Delapierre, M.; Servant, L.; Pope, M.A.; Goldthorpe, I.A. The Joule heating problem in silver nanowire transparent electrodes. *Nanotechnology* **2017**, *28*, 425703. [CrossRef]
18. Lin, X.; Luo, H.; Jia, X.; Wang, J.; Zhou, J.; Jiang, Z.; Pan, L.; Huang, S.; Chen, X. Efficient and ultraviolet durable inverted polymer solar cells using thermal stable GZO-AgTi-GZO multilayers as a transparent electrode. *Org. Electron.* **2016**, *39*, 177–183. [CrossRef]
19. Zhao, Z.; Alford, T.L. The optimal TiO2/Ag/TiO2 electrode for organic solar cell application with high device-specific Haacke figure of merit. *Sol. Energy Mater. Sol. Cells* **2016**, *157*, 599–603. [CrossRef]
20. Lee, S.-M.; Koo, H.-W.; Kim, T.-W.; Kim, H.-K. Asymmetric ITO/Ag/ZTO and ZTO/Ag/ITO anodes prepared by roll-to-roll sputtering for flexible organic light-emitting diodes. *Surf. Coat. Tech.* **2018**, *343*, 115–120. [CrossRef]
21. Axelevitch, A.; Gorenstein, B.; Golan, G. Investigation of Optical Transmission in Thin Metal Films. *Physics Procedia* **2012**, *32*, 1–13. [CrossRef]
22. Yamamoto, N.; Osone, S.; Makino, H.; Yamamoto, T. Influence of Alkaline Chemicals on Electrical and Optical Characteristics of Ga-Doped ZnO Transparent Thin Films. *ECS Trans.* **2011**, *33*, 29–36. [CrossRef]
23. Abduev, A.K.; Akhmedov, A.K.; Asvarov, A.S.; Abdullaev, A.A.; Sulyanov, S.N. Effect of growth temperature on properties of transparent conducting gallium-doped ZnO films. *Semiconductors* **2010**, *44*, 32–36. [CrossRef]
24. Kim, D.; Cho, K.; Kim, H. Thermally evaporated indium-free, transparent, flexible SnO2/AgPdCu/SnO2 electrodes for flexible and transparent thin film heaters. *Sci. Rep.* **2017**, *7*, 2550. [CrossRef] [PubMed]
25. Nakanishi, Y.; Miyake, A.; Kominami, H.; Aoki, T.; Hatanaka, Y.; Shimaoka, G. Preparation of ZnO thin films for high-resolution field emission display by electron beam evaporation. *Appl. Surf. Sci.* **1999**, *142*, 233–236. [CrossRef]
26. Ravichandran, K.; Subha, K.; Manivasaham, A.; Sridharan, M.; Arund, T.; Ravidhas, C. Fabrication of a novel low-cost triple layer system (TaZO/Ag/TaZO) with an enhanced quality factor for transparent electrode applications. *RSC Adv.* **2016**, *6*, 63314–63324. [CrossRef]
27. Rana, A.S.; Chang, S.B.; Chae, H.U.; Kim, H.S. Structural, optical, electrical and morphological properties of different concentration sol-gel ZnO seeds and consanguineous ZnO nanostructured growth dependence on seeds. *J. Alloys Compd.* **2017**, *729*, 571–582. [CrossRef]
28. El Hajj, A.; Lucas, B.; Chakaroun, M.; Antony, R.; Ratier, B.; Aldissi, M. Optimization of ZnO/Ag/ZnO multilayer electrodes obtained by Ion Beam Sputtering for optoelectronic devices. *Thin Solid Films* **2012**, *520*, 4666–4668. [CrossRef]
29. Duta, M.; Anastasescu, M.; Calderon-Moreno, J.M.; Predoana, L.; Preda, S.; Nicolescu, M.; Stroescu, H.; Bratan, V.; Dascalu, I.; Aperathitis, E.; et al. Sol–gel versus sputtering indium tin oxide films as transparent conducting oxide materials. *J. Mater. Sci.: Mater. Electron.* **2016**, *27*, 4913–4922. [CrossRef]
30. Tabassum, S.; Yamasue, E.; Okumura, H.; Ishihara, K.N. Sol–gel and rf sputtered AZO thin films: Analysis of oxidation kinetics in harsh environment. *J. Mater. Sci: Mater. Electron.* **2014**, *25*, 4883–4888. [CrossRef]
31. Kim, M.Y.; Son, K.T.; Lim, D. Effect of an Ag Insertion Layer on the Optical and Electrical Properties of Ga Doped Zinc Oxide Films. *J. Nanosci. Nanotechnol.* **2015**, *15*, 2478–2481. [CrossRef] [PubMed]
32. Stefaniuk, T.; Wróbel, P.; Górecka, E.; Szoplik, T. Optimum deposition conditions of ultrasmooth silver nanolayers. *Nanoscale Res. Lett.* **2014**, *9*, 153. [CrossRef] [PubMed]
33. Alvarez, R.; Gonzalez, J.C.; Espinos, J.P.; Gonzalez-Elipe, A.R.; Cueva, A.; Villuendas, F. Growth of silver on ZnO and SnO2 thin films intended for low emissivity applications. *Appl. Surf. Sci.* **2013**, *268*, 507–515. [CrossRef]
34. Arbab, M. The base layer effect on the d.c. conductivity and structure of dc magnetron sputtered thin films of silver. *Thin Solid Films* **2001**, *381*, 15–31. [CrossRef]
35. Yamada, T.; Nebiki, T.; Kishimoto, S.; Makino, H.; Awai, K.; Narusawa, T.; Yamamoto, T. Dependences of structural and electrical properties on thickness of polycrystalline Ga-doped ZnO thin films prepared by reactive plasma deposition. *Superlattices Microstruct.* **2007**, *42*, 68–73. [CrossRef]

36. Abduev, A.; Akmedov, A.; Asvarov, A.; Chiolerio, A. A Revised Growth Model for Transparent Conducting Ga Doped ZnO Films: Improving Crystallinity by Means of Buffer Layers. *Plasma Process. Polym.* **2015**, *12*, 725–733. [CrossRef]
37. Formica, N.; Ghosh, D.S.; Carrilero, A.; Chen, T.L.; Simpson, R.E.; Pruneri, V. Ultrastable and atomically smooth ultra-thin silver films grown on a copper seed layer. *ACS Appl. Mater. Interfaces* **2013**, *5*, 3048–3053. [CrossRef]
38. Schubert, S.; Meiss, J.; Müller-Meskamp, L.; Leo, K. Improvement of Transparent Metal Top Electrodes for Organic Solar Cells by Introducing a High Surface Energy Seed Layer. *Adv. Energy Mater.* **2013**, *3*, 438–443. [CrossRef]
39. Yuan, Z.S.; Wu, C.C.; Tzou, W.C.; Yang, C.F.; Chen, Y.H. Investigation of high transparent and conductivity of IGZO/Ag/IGZO sandwich structures deposited by sputtering method. *Vacuum* **2019**, *165*, 305–310. [CrossRef]
40. Cheng, C.H.; Ting, J.-M. Transparent conducting GZO, Pt/GZO, and GZO/Pt/GZO thin films. *Thin Solid Films* **2007**, *516*, 203–207. [CrossRef]
41. Yang, H.; Shin, S.; Park, J.; Ham, G.; Oh, J.; Jeon, H. Effect of Au interlayer thickness on the structural, electrical, and optical properties of GZO/Au/GZO multilayers. *Curr. Appl. Phys.* **2014**, *14*, 1331–1334. [CrossRef]
42. Zhang, Q.; Zhao, Y.; Jia, Z.; Qin, Z.; Chu, L.; Yang, J.; Zhang, J.; Huang, W.; Li, X. High Stable, Transparent and Conductive ZnO/Ag/ZnO Nanofilm Electrodes on Rigid/Flexible Substrates. *Energies* **2016**, *9*, 443. [CrossRef]
43. Park, H.-K.; Jeong, J.-A.; Park, Y.-S.; Na, S.-I.; Kim, D.-Y.; Kim, H.-K. Room-Temperature Indium-Free Ga:ZnO/Ag/Ga:ZnO Multilayer Electrode for Organic Solar Cell Applications Electrochem. *Solid-State Lett.* **2009**, *12*, H309. [CrossRef]
44. Wu, H.-W.; Yang, R.-Y.; Hsiung, C.-M.; Chu, C.-H. Influence of Ag thickness of aluminum-doped ZnO/Ag/aluminum-doped ZnO thin films. *Thin Solid Films* **2012**, *520*, 7147–7152. [CrossRef]
45. Sahu, D.R.; Huang, J.-L. High quality transparent conductive ZnO/Ag/ZO multilayer films deposited at room temperature. *Thin Solid Films* **2006**, *515*, 876–879. [CrossRef]
46. Yu, S.; Li, L.; Lyu, X.; Zhang, W. Preparation and investigation of nano-thick FTO/Ag/FTO multilayer transparent electrodes with high fgure of merit. *Sci. Rep.* **2016**, *6*, 20399. [CrossRef]
47. Haacke, G. New fgure of merit for transparent conductors. *J. Appl. Phys.* **1976**, *47*, 4086–4089. [CrossRef]
48. Kato, K.; Omoto, H.; Takamatsu, A. Optimum structure of metal oxide under-layer used in Ag-based multilayer. *Vacuum* **2008**, *84*, 606–609. [CrossRef]
49. Andoa, E.; Miyazaki, M. Moisture degradation mechanism of silver-based low-emissivity coatings. *Thin Solid Films* **1999**, *351*, 308–312. [CrossRef]

© 2020 by the authors. Licensee MDPI, Basel, Switzerland. This article is an open access article distributed under the terms and conditions of the Creative Commons Attribution (CC BY) license (http://creativecommons.org/licenses/by/4.0/).

Article

Optoelectronic Properties of Ti-doped SnO₂ Thin Films Processed under Different Annealing Temperatures

Chi-Fan Liu [1], Chun-Hsien Kuo [2], Tao-Hsing Chen [1,*] and Yu-Sheng Huang [1]

[1] Department of Mechanical Engineering, National Kaohsiung University of Science and Technology, Kaohsiung 80778, Taiwan; i108142101@nkust.edu.tw (C.-F.L.); isu10207025a@gmail.com (Y.-S.H.)
[2] Department of Mold and Die Engineering, National Kaohsiung University of Science and Technology, Kaohsiung 80778, Taiwan; chkuo@nkust.edu.tw
* Correspondence: thchen@nkust.edu.tw; Tel.: +886-7-3814526-15330

Received: 20 March 2020; Accepted: 13 April 2020; Published: 16 April 2020

Abstract: Ti-doped SnO_2 transparent conductive oxide (TCO) thin films are deposited on glass substrates using a radio frequency (RF) magnetron sputtering system and then are annealed at temperatures in the range of 200–500 °C for 30 min. The effects of the annealing temperature on the structural properties, surface roughness, electrical properties, and optical transmittance of the thin films are then systematically explored. The results show that a higher annealing temperature results in lower surface roughness and larger crystal size. Moreover, an annealing temperature of 300 °C leads to the minimum electrical resistivity of 5.65×10^{-3} Ω·cm. The mean optical transmittance increases with an increase in temperature and achieves a maximum value of 74.2% at an annealing temperature of 500 °C. Overall, the highest figure of merit (Φ_{TC}) (3.99×10^{-4} Ω^{-1}) is obtained at an annealing temperature of 500 °C.

Keywords: SnO_2; Ti-doped; annealing temperature; electrical resistivity; transmittance

1. Introduction

Transparent conductive oxide (TCO) thin films possess excellent conductivity and optical transmittance in the visible and near-infrared regions, and are thus applied in many photoelectric components nowadays, including solar cells [1,2], organic light-emitting diodes [3,4], thin-film transistors [5,6], photovoltaic batteries [7–9], electrochromic devices [10–12], and tablet displays [13–16]. Metallic films are generally opaque in the visible light range. However, for film thicknesses of less than 100 Å, visible light is transmitted through the film, while infrared (IR) light is reflected. Moreover, for metals such as In_2O_3, ZnO, SnO_2, TiO_2, and CdO with energy gaps of 3 eV or more, the film also has excellent semiconducting properties [17].

The literature contains many studies on the optoelectronic properties of metallic films [18–20]. In addition, various authors have investigated the properties of three-layer TCO films with oxide/metal/oxide or metallic oxide/metal/metallic oxide structures [21]. The results have shown that such films not only suppress the reflection from the metallic layer in the visible light range but also produce a transmittance effect [22,23]. Consequently, the TCO thin films are used in solar cells, gas sensors, LCD displays, etc.

Among the various metal oxides in common use nowadays, SnO_2 has poorer electrical properties than ITO, but a superior photoelectric performance in the IR region. Furthermore, SnO_2 has good chemical and thermal stability and is also amenable to surface modification in order to expand its working wavelength range. As a result, SnO_2 conductive films are widely used for such applications as gas sensors, solar energy battery electrodes, low-radiation glasses, etc. [24,25]. However, SnO_2 films are less easily used in tablet display applications due to their high electrical resistance and poor etching effect.

Accordingly, the present study explores the feasibility of improving the optical and electrical properties of SnO_2 thin films by doping the films with Ti. Note that Ti is deliberately selected as the dopant material here, since it has a maximum chemical valence of +4 [26], where the radius of Ti^{4+} is 0.0605 nm, while that of Sn^{4+} is 0.069 nm. Due to the similarity of the ion radii, the Ti^{4+} ions readily replace the Sn^{4+} ions in the crystal lattice of the SnO_2 and hence modify its electrical and optical behavior. The Ti:SnO_2 films are deposited on glass substrates using a radio frequency (RF) magnetron sputtering system and then are annealed at various temperatures in the range of 200–500 °C to prompt the diffusion of the Ti atoms into the SnO_2 layer. The optoelectronic properties of the films are then systematically explored in order to determine the annealing temperature, which results in the optimal tradeoff between the electrical and optical properties of the film, respectively.

2. Experimental Procedure

The glass plate was purchased from Corning company (Corning, NY, USA) and cut into pieces the size of 25 mm × 25 mm × 7 mm (length × width × thickness) using a diamond saw. The substrates were cleaned sequentially in deionized (DI) water, acetone, and IPA (isopropanol), and DI water once again in order to remove any pollutants, residual solvents, or nonorganic components from the substrate surface. The substrates were then dried in an oven at 90 °C until the water was completely vaporized. The thin films were prepared using a sputtering target (two-inch diameter) composed of SnO_2(95%) and TiO_2(5%). The Ti:SnO_2 films were then deposited on the glass substrates using an RF sputtering system with a sputtering power of 60 W, an argon gas flow rate of 29 sccm, an oxygen flow rate of 1 sccm, and a bias of 7.5 mTorr. The sputtering process was performed without substrate heating. The purpose of this experiment was to study the impact of different process parameters and conditions on the characteristics of a transparent conductive film. Following the deposition process, the Ti:SnO_2 films were annealed at temperatures of 200, 300, 400, and 500 °C for 30 min. The structures of the annealed thin films were examined by X-ray diffraction (XRD, Bruker, Billerica, MA, USA). In addition, the surface roughness and crystal size were determined by atomic force microscopy (AFM, NTMDT-AFM, Bruker, Billerica, MA, USA) and scanning electron microscopy (SEM, JEOL JSM-7000F, JEOL, Kyoto, Janpan), respectively. The photoelectric property data consisted of light transmittance, resistivity, carrier concentration by spectrophotometer (UV Spectrophotometer, Hitachi 2900, Hitachi, Tokyo, Japan), and a Hall measuring instrument(AHM-800B, Advnaced Design Technology, Taipei, Taiwan), respectively. In this paper, the measurement for each condition was performed six times to confirm the data.

3. Results and Discussion

3.1. Effects of Annealing Temperature on Ti:SnO_2 Film Thickness

Table 1 shows the thickness of the various Ti:SnO_2 films, measured by alpha–step profilometer (KLA-Tencor, Milpitas, CA, USA). In general, it is noted that while the annealing temperature has no significant effect on the film thickness, the thickness increases slightly in the film annealed at 200 °C but then reduces progressively as the annealing temperature is further increased to 500 °C.

Table 1. Effects of annealing temperature on thickness of Ti:SnO_2 films.

Annealing Temperature (°C)	Thickness (nm)
as-deposited	88.2 ± 2.0
200	93 ± 2.0
300	90 ± 2.0
400	89.4 ± 2.0
500	88.6 ± 2.0

3.2. Effects of Annealing Temperature on Structural Properties of Ti:SnO$_2$ Films

Figure 1 shows the XRD patterns of the as-deposited and annealed Ti:SnO$_2$ thin films. As the annealing temperature increases, prominent peaks are observed at 26.6°, 33.8°, and 51.7° corresponding to (110), (101), and (211) phases, respectively. The (101) phase is a combined SnO and SnO$_2$ phase [27]; some of the Sn and Ti atoms are replaced by a diffused process following the annealing process [28,29]. Notably, diffraction peaks are very small in the as-deposited film or the film annealed at 200 °C. However, as the annealing temperature increases to 300 °C, the crystalline phase appears more within the film. The crystalline structure becomes increasingly pronounced as the annealing temperature increases to 500 °C, and hence has a significant effect on the electrical and optical properties, as described in the following sections.

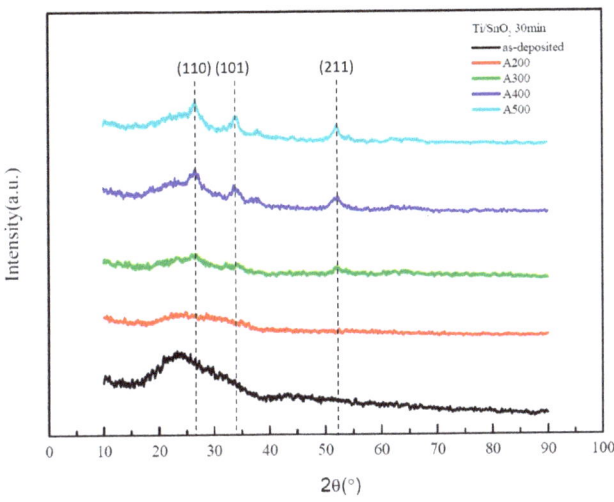

Figure 1. X-ray diffraction (XRD) patterns of as-deposited and annealed Ti:SnO$_2$ films.

3.3. Effects of Annealing Temperature on Electrical Resistivity

Figure 2 shows the electrical properties of the as-deposited and annealed Ti:SnO$_2$ films. Previous studies have shown that the TCO transmission mechanism is governed mainly by element doping and oxygen vacancies [29,30]. For the Ti:SnO$_2$ thin films considered in the present study, the oxygen vacancy contributes two free electrons, and therefore dominates the transmission mechanism. Although the Sn$^+$ atom also provides a free electron, it cannot be activated as effectively as the oxygen vacancy because the carrier concentration is primarily controlled by the oxygen vacancy. As described above, the Ti:SnO$_2$ film has a small crystal structure in the as-deposited condition and under an annealing temperature of 200 °C. However, as the annealing temperature is increased to 300 °C, the film has a low resistivity of 5.65×10^{-3} Ω·cm as a result of the high carrier concentration. Meantime, as the annealing temperature is increased beyond 300 °C, the SnO and SnO$_2$ combined phase are gradually formed, causing the carrier concentration to decrease and the resistivity to increase.

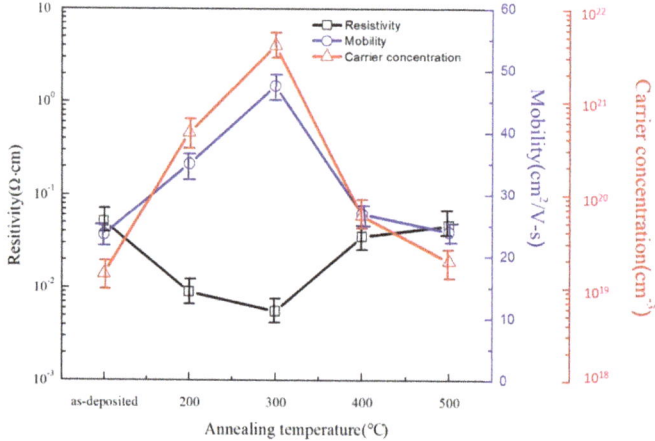

Figure 2. Electrical properties of as-deposited and annealed Ti:SnO$_2$ films.

3.4. Transmittance

Figure 3 shows the optical transmittance properties of the various Ti:SnO$_2$ thin films. For the as-deposited film, the mean transmittance has a low value of approximately 58% due to the poor effect of the replaced Sn for the Ti atom. However, the transmittance improves significantly in the annealed samples, particularly in those annealed at temperatures of 300 °C or more. In conventional ITO films, the optical energy gap theoretically increases as the carrier concentration increases, since the Fermi level moves into the conduction band and the electrons on the valence band are forced to jump to the conduction band, thereby requiring more energy and resulting in the so-called Burstein–Moss effect [31,32]. However, the optical energy gap rises with a decreasing carrier concentration. Such a phenomenon may be due to an interaction effect between ion compounds. For example, Zn^{2+} and Sn^{4+} ions coexist in IZTO films and trigger the generation of a donor-acceptor pair, which reduces the energy gap and mitigates the Burstein–Moss effect. For the present Ti:SnO$_2$ films, the carrier concentration decreases following annealing at temperatures higher than 200 °C (see Figure 2). However, the energy gap and mean transmittance both increase (see Table 2 and Figure 3, respectively). For an annealing temperature of 200 °C, the improvement in the transmittance is very modest (i.e., from around 58% for the as-deposited sample to approximately 60% for the annealed sample). However, for an annealing temperature of 300 °C, the film undergoes a transformation from a homogenous crystalline structure and the mean transmittance improves to almost 75%. Furthermore, as the annealing temperature increases, the transformation toward a crystalline structure becomes more complete (see Figure 1) and hence the mean transmittance increases. Thus, the film annealed at a temperature of 500 °C shows the maximum mean transmittance of approximately 74.2%.

Table 2. Effects of annealing temperature on energy gap (Eg) of Ti:SnO$_2$ films.

Annealing Temperature (°C)	Eg (eV)
as-deposited	2.95
200	2.88
300	3.13
400	3.21
500	3.28

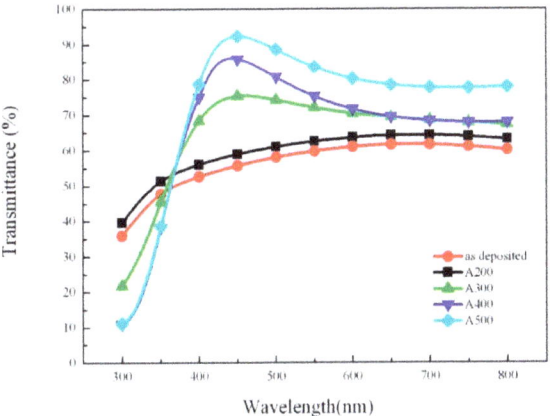

Figure 3. Optical transmittance of as-deposited and annealed Ti:SnO$_2$ films.

Figure 4 shows the relationship between the optical absorption coefficient (α) and photon energy($h\nu$) for the Ti:SnO$_2$ film. The optical band gap (E_g) is calculated as follows with the equation [33,34]:

$$\alpha h\nu = A(h\nu - E_g)^{1/2} \tag{1}$$

where α is the absorption coefficient, ν is the frequency of incident light, h is the Planck's constant, and A is constant. The optical band gap is extrapolating the straight-line portion of the plot to the energy axis. Table 2 shows the calculated values of the optical band gap for the present Ti:SnO$_2$ thin films. Furthermore, an Eg value greater than 3 eV is regarded as excellent. Referring to Table 2, the Eg value of the present Ti:SnO$_2$ films increases with an increase in annealing temperature and is equal to 3.28 eV at an annealing temperature of 500 °C. Moreover, an Eg value greater than 3 eV is obtained for all of the films annealed at a temperature of 300 °C or more.

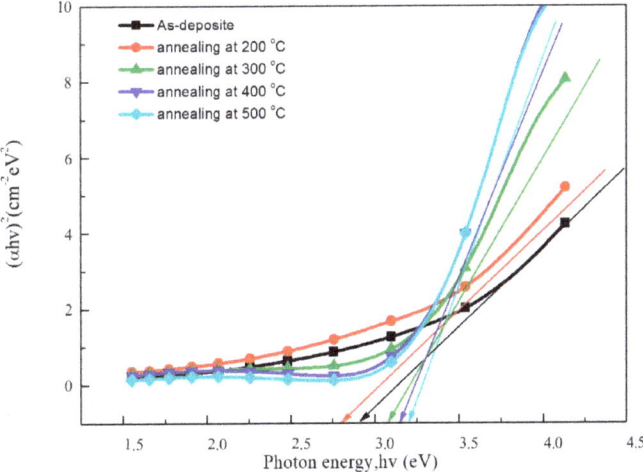

Figure 4. The ($\alpha h\nu$)^2 against photon energy ($h\nu$) of Ti:SnO$_2$ films under different annealing temperatures.

3.5. Surface Feature Analysis

Figures 5 and 6 show AFM and SEM images of the various as-deposited and annealed Ti:SnO$_2$ films. The mean surface roughness values of the films are listed in Table 3. As shown, the as-deposited sample has a surface roughness of 0.31 nm. However, following annealing at 300 °C, the surface roughness falls to a value of around 0.35 nm due to the formation of the crystalline phase. However, as the annealing temperature is further increased, the surface roughness reduces and has a value of just 0.296 nm in the sample processed at the highest annealing temperature of 500 °C. The AFM and SEM images show that the as-deposited Ti:SnO$_2$ film and the film annealed at 300 °C have higher surface roughness. Annealing at a temperature of 300 °C results in high roughness, but after 300 °C the roughness is decreased.

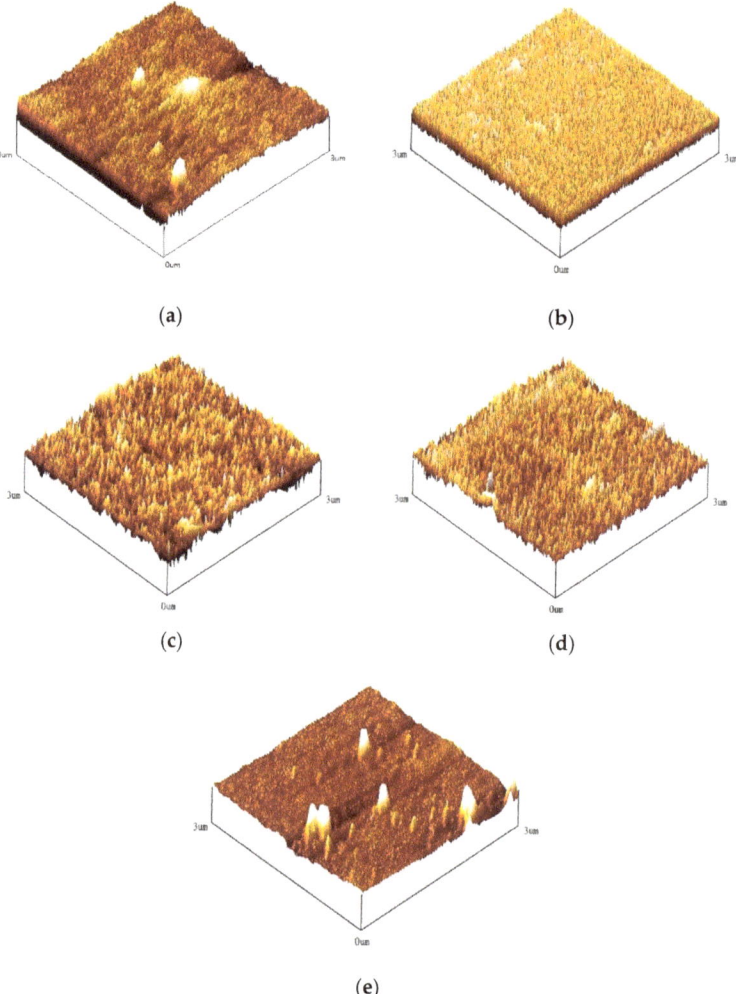

Figure 5. Atomic force microscopy (AFM) images of as-deposited and annealed Ti:SnO$_2$ films: (**a**) as-deposited, (**b**) annealed at 200 °C, (**c**) annealed at 300 °C, (**d**) annealed at 400 °C, and (**e**) annealed at 500 °C.

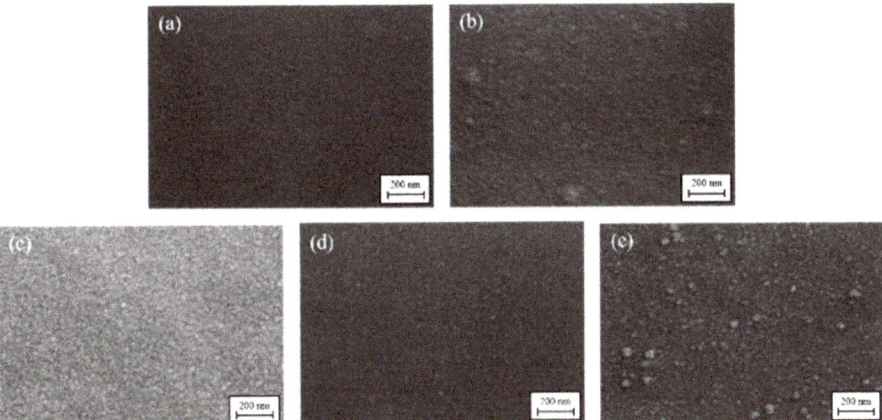

Figure 6. SEM images of as-deposited and annealed Ti:SnO$_2$ films: (**a**) as-deposited, (**b**) annealed at 200 °C, (**c**) annealed at 300 °C, (**d**) annealed at 400 °C, and (**e**) annealed at 500 °C.

Table 3. Effects of annealing temperature on surface roughness of Ti:SnO$_2$ films.

Annealing Temperature (°C)	Ra (nm)
as-deposited	3.10 ± 0.02
200	3.23 ± 0.02
300	3.50 ± 0.02
400	3.03 ± 0.02
500	2.96 ± 0.02

3.6. Effects of Annealing Temperature on Crystal Size

The full width at half maximum (FWHM) values of the peaks in the XRD patterns can be derived from the following Scherrer formula [35]:

$$D = 0.9 \times \lambda/\beta\cos\theta \qquad (2)$$

where D is the grain size, β is the XRD peak FWHM, λ is the wavelength of the incident light, and θ is the diffraction angle of the incident light. In the XRD process, λ and θ have constant values. Consequently, the grain size, D, and FWHM, β, are inversely related. (Cullity and Stock 2001). Figure 7 shows the FWHM and crystal grain size values of the present Ti:SnO$_2$ films. Note that the as-deposited film has a small crystal structure, and hence the FWHM and crystal size values are also calculated carefully. However, for an annealing temperature of 300 °C, the Ti:SnO$_2$ film has a crystalline structure with a grain size of around 14.89 nm. As the annealing temperature is increased, the crystalline structure becomes more pronounced. Consequently, the grain size decreases, while the FWHM increases. For the maximum annealing temperature of 500 °C, the grain size is equal to approximately 11.56 nm, while the FWHM increases to 0.8.

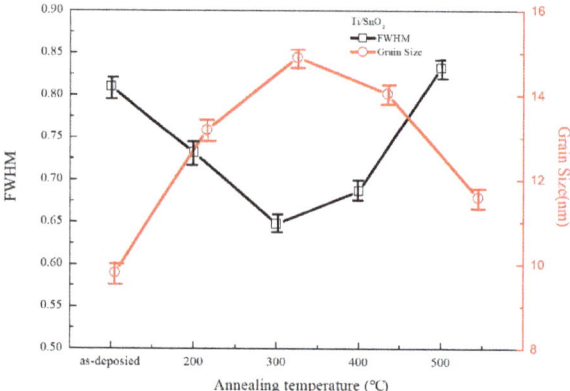

Figure 7. The full width at half maximum (FWHM) and grain size of as-deposited and annealed Ti:SnO$_2$ films.

3.7. Figure of Merit(Φ_{TC})

The figure of merit (Φ_{TC}) is an important factor used to evaluate the performance of TCO films from the relationship between transmittance and electrical properties. The figure of merit (Φ_{TC}) is defined as [36] as follows:

$$\Phi_{TC} = T^{10}/R_{sh} \qquad (3)$$

where T is the average optical transmittance, and R_{sh} is the sheet resistance of the films. Table 4 shows the figure of merit(Φ_{TC}) results for the as-deposited and annealed Ti:SnO$_2$ thin films. As shown, the optimal Φ_{TC} (3.99 × 10^{-4} Ω^{-1}) is obtained at an annealing temperature of 500 °C. According to the figure, we see that because of an insignificant difference in electrical properties, it has led to a more obvious influence on quality elements from mean optical transmittance, and the optimal mean transmittance is seen at 500 °C. Therefore, optimum quality elements can be achieved at 500 °C.

Table 4. The figure of merit of Ti/SnO$_2$ (Φ_{TC} (Ω^{-1})).

Annealing Temperature (°C)	Ti/SnO$_2$
as-deposited	1.61 × 10^{-5}
200	2.79 × 10^{-5}
300	1.99 × 10^{-4}
400	2.27 × 10^{-4}
500	3.99 × 10^{-4}

3.8. Comparison of Other Methods to the Ti-Doped SnO$_2$ Method

Table 5 shows some literature about the Ti-doped SnO$_2$ thin film under different methods (e.g., sol-gel and ultrasonic spray). We listed the optical and electrical properties in Table 5 from the literature and this study. From Table 5, it can be seen that the RF sputter method has some better optical and electrical properties than sol-gel. However, the cost of the sputtering method is more expansive.

Table 5. A comparison of the values of film properties in this study with other methods.

Method	Transmittance (%)	Resistivity ($\Omega \cdot$cm)	Band Gap(eV)
Rf-Sputter (this study, the doped content of Ti is at 5 at %)	Average is 74.2% (annealing at 500 °C, maximum is 92%)	5.65×10^{-3} (annealing at 300 °C)	3.21 (annealing at 500 °C)
Ultrasonic spray [37]	Maximum 83% (the doped content of Ti is at 4 at %)	7.64×10^{-3} (the doped content of Ti is at 4 at %)	3.91 (the doped content of Ti is at 4 at %)
Sol-gel [38]	none	62.5 (calcined at 1000 °C)	none
Sol-gel [39]	Maximum 74% (the doped content of Ti is at 5 at %)	None	3.67 (the doped content of Ti is at 5 at %)

4. Conclusions

This study has examined the electrical and optical properties of Ti:SnO$_2$ thin films deposited on glass substrates and then annealed at temperatures ranging from 200–500 °C. The experimental results have shown that the thickness of the Ti:SnO$_2$ films is insensitive to the annealing temperature. However, as the annealing temperature increases, strong peaks in the XRD patterns emerge corresponding to (110), (101), and (211) phases. Hence, it is inferred that the films increase a well-crystalline structure at higher annealing temperatures. The Ti:SnO$_2$ film annealed at the lowest temperature of 300 °C shows both the minimum resistivity of 5.65×10^{-3} $\Omega \cdot$cm. The energy gap and optical transmittance both increase with increasing annealing temperature and have values of 3.28 eV and 74.2% at an annealing temperature of 500 °C. The AFM results show that for the samples annealed at temperatures of more than 300 °C, the mean surface roughness reduces with an increase in annealing temperature. The SEM observations suggest that the lower surface roughness is the result of larger grain size. In particular, the grain size decreases from 14.89 nm in the film annealed at 300 °C to 11.56 nm in the film annealed at 500 °C. The Ti:SnO$_2$ film annealed at a temperature of 500 °C shows the highest Φ_{TC} of 3.99×10^{-4} Ω^{-1}. The characterization results have suggested that the optimal performance of this film is due to optical transmittance.

Author Contributions: Data curation—formal analysis, C.-F.L. and Y.-S.H.; Writing—Original draft preparation—methodology—investigation T.-H.C.; Review and editing—investigation C.-H.K. All authors have read and agreed to the published version of the manuscript.

Funding: This research was funded by the Ministry of Science and Technology, Taipei, Taiwan, under Grant No. 106-2628-E-992 -302 -MY3.

Acknowledgments: The authors gratefully acknowledge the financial support provided to this study by the Ministry of Science and Technology.

Conflicts of Interest: The authors declare no conflict of interest.

References

1. Fleischer, K.; Arca, E.; Shvets, I.V. Improving solar cell efficiency with optically optimised TCO layers. *Sol. Energy Mater. Sol. Cells* **2012**, *101*, 262–269. [CrossRef]
2. Sun, W.; Wang, S.; Li, S.; Miao, X.; Zhu, Y.; Du, C.; Ma, R.; Wang, C. Reactive-Sputtered Prepared Tin Oxide Thin Film as an Electron Transport Layer for Planar Perovskite Solar Cells. *Coatings* **2019**, *9*, 320. [CrossRef]
3. Oh, M.; Seo, I. Enhanced performance of GaN-based green light-emitting diodes with gallium-doped ZnO transparent conducting oxide. *J. Electron. Mater.* **2014**, *43*, 1232–1236. [CrossRef]
4. Najafi, N.; Rozati, S.M. Structural and Electrical Properties of SnO$_2$:F Thin Films Prepared by Chemical Vapor Deposition Method. *Acta Phys. Pol. A* **2017**, *131*, 222–225. [CrossRef]
5. Jang, B.; Kim, T.; Lee, S.; Lee, W.Y.; Kang, H.; Cho, C.S.; Jang, J. High Performance Ultrathin SnO$_2$ Thin-Film Transistors by Sol–Gel Method. *IEEE Electron Device Lett.* **2018**, *39*, 1179–1182. [CrossRef]
6. Presley, R.E.; Munsee, C.L.; Park, C.H.; Hong, D.; Wager, J.F.; Keszler, D.A. Tin oxide transparent thin-film transistors. *J. Phys D Appl. Phys.* **2004**, *37*, 2810–2813. [CrossRef]
7. Kim, C.; Noh, M.; Choi, M.; Cho, J.; Park, B. Critical size of a nano SnO$_2$ electrode for Li secondary battery. *Chem. Mater.* **2005**, *17*, 3297–3301. [CrossRef]

8. Riveros, R.; Romero, E.; Gordillo, G. Synthesis and Characterization of Highly Transparent and Conductive SnO$_2$:F and In$_2$O$_3$:Sn thin Films Deposited by Spray Pyrolysis. *Braz. J. Phys.* **2006**, *36*, 1042–1045. [CrossRef]
9. Haider, A.J.; Mohammed, A.J.; Shaker, S.S.; Yahya, K.Z.; Haider, M.J. Sensing Characteristics of Nanostructured SnO$_2$ Thin Films as Glucose sensor. *Energy Procedia* **2017**, *119*, 473–481. [CrossRef]
10. Patil, P.S.; Sadale, S.B.; Mujawar, S.H.; Shinde, P.S.; Chigare, P.S. Synthesis of electrochromic tin oxide thin films with faster response by spray pyrolysis. *Appl. Sur. Sci.* **2007**, *253*, 8560–8567. [CrossRef]
11. Chang, J.Y.; Chen, Y.C.; Wang, C.M.; Chen, Y.W. Electrochromic Properties of Li-Doped NiO Films Prepared by RF Magnetron Sputtering. *Coatings* **2020**, *10*, 87. [CrossRef]
12. Vernardou, D. Using an Atmospheric Pressure Chemical Vapor Deposition Process for the Development of V$_2$O$_5$ as an Electrochromic Material. *Coatings* **2017**, *7*, 24. [CrossRef]
13. Isono, T.; Fukuda, T.; Nakagawa, K.; Usui, R.; Satoh, R.; Morinaga, E.; Mihara, Y. Highly conductive SnO2thin films for flat-panel displays. *J. Soc. Inf. Disp.* **2007**, *15*, 161–166. [CrossRef]
14. Patel, P.; Karmakar, A.; Jariwal, C.; Ruparelia, J.P. Preparation and Characterization of SnO$_2$ Thin Film Coating using rf-Plasma Enhanced Reactive Thermal Evaporation. *Procedia Eng.* **2013**, *51*, 473–479. [CrossRef]
15. Haider, A.J.; Shaker, S.S.; Mohammed, A.H. A Study of Morphological, Optical and Gas Sensing Properties for Pure and Ag Doped SnO$_2$ Prepared by Pulsed Laser Deposition (PLD). *Energy Procedia* **2013**, *36*, 776–787. [CrossRef]
16. Lee, S.H.; Kwon, K.; Kim, K.; Yoon, J.S.; Choi, D.S.; Yoo, Y.; Kim, C.; Kang, S.; Kim, J.H. Electrical, Structural, Optical, and Adhesive Characteristics of Aluminum-Doped Tin Oxide Thin Films for Transparent Flexible Thin-Film Transistor Applications. *Materials* **2019**, *12*, 137. [CrossRef]
17. Shanthi, E.; Banerjee, A.; Dutta, V.; Chopra, K.L. Electrical and optical properties of tin oxide films doped with F and (Sb + F). *J. Appl. Phys.* **1982**, *53*, 1615–1621. [CrossRef]
18. Chen, T.H.; Chen, T.Y. Effects of Annealing Temperature on Properties of Ti-Ga–Doped ZnO Films Deposited on Flexible Substrates. *Nanomaterials* **2015**, *5*, 1831. [CrossRef]
19. Chen, T.H.; Jiang, B.L. Optical and electronic properties of Mo:ZnO thin films deposited using RF magnetron sputtering with different process parameters. *Opt. Quant. Electron* **2016**, *48*, 77. [CrossRef]
20. Chen, T.H.; Jiang, B.L.; Huang, C.T. The Optical and Electrical Properties of MZO Transparent Conductive Thin Films on Flexible Substrate. *Smart Sci.* **2017**, *5*, 53–60. [CrossRef]
21. Chen, T.H.; Su, H.T. Effect of Annealing Temperature on Optical and Electrical Properties of ZnO/Ag/ZnO Multilayer Films for Photosensor. *Sens. Mater.* **2018**, *30*, 2541–2547. [CrossRef]
22. Fan, J.C.C.; Bachner, F.J.; Foley, G.H.; Zavracky, P.M. Transparent heat-mirror films of TiO$_2$/Ag/TiO$_2$ for solar energy collection and radiation insulation. *Appl. Phys. Lett.* **1974**, *25*, 693–695. [CrossRef]
23. Vidor, F.F.; Meyers, T.; Hilleringmann, U. Circuits Using ZnO Nanoparticle Based Thin-Film Transistors for Flexible Electronic Applications. *Nanomaterials* **2016**, *6*, 154. [CrossRef] [PubMed]
24. Lee, J.J.; Ha, J.Y.; Choi, W.K.; Cho, Y.S.; Choi, J.W. Doped SnO$_2$ Transparent Conductive Multilayer Thin Films Explored by Continuous Composition Spread. *ACS Comb. Sci.* **2015**, *17*, 247–252. [CrossRef]
25. Esro, M.; Georgakopoulos, S.; Lu, H.; Vourlias, G.; Krier, A.; Milne, W.I.; Gillin, W.P.; Adamopoulos, G. Solution processed SnO$_2$:Sb transparent conductive oxide as an alternative to indium tin oxide for applications in organic light emitting diodes. *J. Mater. Chem. C* **2016**, *4*, 3563–3570. [CrossRef]
26. Liu, W.S.; Hsieh, W.T.; Chen, S.Y.; Huang, C.S. Improvement of CIGS solar cells with high performance transparent conducting Ti-doped GaZnO thin films. *Sol. Energy* **2018**, *174*, 83–96. [CrossRef]
27. Ding, X.; Fang, F.; Jiang, J. Electrical and optical properties of N-doped SnO$_2$ thin films prepared by magnetron sputtering. *Surf. Coat. Technol.* **2013**, *231*, 67–70. [CrossRef]
28. Sousa, M.G.; da Cunha, A.F. Optimization of low temperature RF-magnetron sputtering of indium tin oxide films for solar cell applications. *Appl. Sur. Sci.* **2019**, *484*, 257–264. [CrossRef]
29. Song, S.; Yang, T.; Liu, J.; Xin, Y.; Li, Y.; Han, S. Rapid thermal annealing of ITO films. *Appl. Sur. Sci.* **2011**, *257*, 7061–7064. [CrossRef]
30. Zhu, B.L.; Zhao, X.; Hu, W.C.; Li, T.T.; Wu, J.; Gan, Z.H.; Liu, J.; Zeng, D.W.; Xie, C.S. Structural, electrical, and optical properties of F-doped SnO or SnO$_2$ films prepared by RF reactive magnetron sputtering at different substrate temperatures and O$_2$ fluxes. *J. Alloys Compd.* **2017**, *719*, 429–437. [CrossRef]
31. Wang, Y.; Tang, W. Surface-dependent conductivity, transition type, and energy band structure in amorphous indium tin oxide films. *Solid-State Electron.* **2017**, *138*, 79–83. [CrossRef]

32. Cullity, B.D.; Stock, S.R. *Elements of X-ray Diffraction*; Chapters 5–2; Prentice Hall: Upper Saddle River, NJ, USA, 2001; pp. 167–171.
33. Shan, F.; Kim, B.L.; Liu, G.X.; Liu, Z.F.; Sohn, J.Y.; Lee, W.J.; Shin, B.C.; Yu, Y.S. Blueshift of near band edge emission in Mg doped ZnO thin films and aging. *J. Appl. Phys.* **2004**, *95*, 4772–4776. [CrossRef]
34. Amalathas, A.P.; Alkaisi, M.M. Effects of film thickness and sputtering power on properties of ITO thin films deposited by RF magnetron sputtering without oxygen. *J. Mater. Sci.: Mater. Electron* **2016**, *27*, 11064–11071. [CrossRef]
35. Caglar, M.; Ilican, S.; Caglar, Y. Influence of dopant concentration on the optical properties of ZnO: In films by sol–gel method. *Thin Solid Films* **2009**, *517*, 5023–5028. [CrossRef]
36. Haacke, G. New figure of merit for transparent conductors. *J. Appl. Phys.* **1976**, *47*, 4086–4089. [CrossRef]
37. Khelifi, C.; Attaf, A. Influence of Ti doping on SnO_2 thin films properties prepared by ultrasonic spray technique. *Surf. Interfaces* **2020**, *18*, 100449. [CrossRef]
38. Liu, X.M.; Wu, S.L.; Chu, P.K.; Zheng, J.; Li, S.L. Characteristics of nano Ti-doped SnO_2 powders prepared by sol–gel method. *Mater. Sci. Eng. A* **2006**, *426*, 274–277. [CrossRef]
39. Sakthiraj, K.; Balachandrakumar, K. Influence of Ti addition on the room temperature ferromagnetism of tin oxide (SnO_2) nanocrystal. *J. Man. Magn. Mater.* **2015**, *395*, 205–212. [CrossRef]

© 2020 by the authors. Licensee MDPI, Basel, Switzerland. This article is an open access article distributed under the terms and conditions of the Creative Commons Attribution (CC BY) license (http://creativecommons.org/licenses/by/4.0/).

Article

In-Situ Ellipsometric Study of the Optical Properties of LTL-Doped Thin Film Sensors for Copper(II) Ion Detection

Dervil Cody [1,*], Tsvetanka Babeva [2], Violeta Madjarova [2], Anastasia Kharchenko [3], Sabad-e-Gul [1], Svetlana Mintova [3], Christopher J. Barrett [4] and Izabela Naydenova [1,*]

1. Centre for Industrial and Engineering Optics, School of Physics and Clinical and Optometric Sciences, Technological University Dublin, Kevin Street, D08 NF82 Dublin 8, Ireland; sabadegul@gmail.com
2. Institute for Optical Materials and Technologies, Bulgarian Academy of Sciences, 109, Acad. G. Bontchev Str., P.O. Box 95, 1113 Sofia, Bulgaria; babeva@iomt.bas.bg (T.B.); vmadjarova@iomt.bas.bg (V.M.)
3. Laboratoire Catalyse & Spectrochimie, ENSICAEN, Normandie Université de Caen, CNRS, 6, boulevard du Maréchal Juin, 14050 Caen, France; anastasia.khartchenko@gmail.com (A.K.); svetlana.mintova@ensicaen.fr (S.M.)
4. Department of Chemistry, McGill University, 801 Rue Sherbrooke W, Montréal, QC H3A 0B8, Canada; chris.barrett@mcgill.ca
* Correspondence: dervil.cody@tudublin.ie (D.C.); izabela.naydenova@tudublin.ie (I.N.)

Received: 13 March 2020; Accepted: 21 April 2020; Published: 24 April 2020

Abstract: Optical sensors fabricated in zeolite nanoparticle composite films rely on changes in their optical properties (refractive index, n, and thickness, d) to produce a measurable response in the presence of a target analyte. Here, ellipsometry is used to characterize the changes in optical properties of Linde Type L (LTL) zeolite thin films in the presence of Cu^{2+} ions in solution, with a view to improving the design of optical sensors that involve the change of n and/or d due to the adsorption of Cu^{2+} ions. The suitability of two different ellipsometry techniques (single wavelength and spectroscopic) for the evaluation of changes in n and d of both undoped and zeolite-doped films during exposure to water and Cu^{2+}-containing solutions was investigated. The influence of pre-immersion thermal treatment conditions on sensor response was also studied. Due to the high temporal resolution, single wavelength ellipsometry facilitated the identification of a Cu^{2+} concentration response immediately after Cu^{2+} introduction, indicating that the single wavelength technique is suitable for dynamic studies of sensor–analyte interactions over short time scales. In comparison, spectroscopic ellipsometry produced a robust analysis of absolute changes in film n and d, as well as yielding insight into the net influence of competing and simultaneous changes in n and d inside the zeolite-doped films arising due to water adsorption and the ion exchange of potassium (K^+) cations by copper (Cu^{2+}).

Keywords: optical sensors; optical materials; zeolites; ellipsometry; single wavelength ellipsometry; spectroscopic ellipsometry

---

1. Introduction

Sensors are being developed for every possible aspect of modern life, ranging from the detection of air pollutants [1] and food contaminants [2], monitoring health biomarkers [3,4], and even assisting in the detection of extra-terrestrial life [5]. There are many characteristics of an effective sensor, including high sensitivity, selectivity, high signal-to-noise ratio, fast response time, and reliability/stability. Sensors should ideally also be manufacturable at relatively low cost and have a reasonable shelf life that makes them cost-effective to use.

Optical diffraction grating-based sensors respond to an analyte or environmental stressor via a change in their optical properties, namely grating refractive index, n, refractive index modulation, Δn, and/or thickness, d. In the case of surface relief grating configuration sensors, Δn is the difference in refractive index, n, between the surface relief grating material and the surrounding medium. For volume grating configuration sensors, Δn is the difference in the refractive index, n, between illuminated and non-illuminated regions inside the grating. In the case of transmission-mode diffraction grating-based sensors illuminated with a probe beam (with incident intensity I_o), any change in the value of Δn or d will vary the phase difference, φ, between the beams propagating along the zero (I_t) direction, and the higher orders (I_d) of diffraction from the grating. For thin gratings operating in the Raman-Nath regime [6], the diffraction efficiency, η, can be related to φ via:

$$\eta = J_m^2\left(\frac{\varphi}{2}\right) = \frac{I_d}{I_o} \qquad (1)$$

where J_m is the Bessel function of the order m, and φ is given by:

$$\varphi = \frac{2\pi \Delta n d}{\lambda_r \cos \theta_B} \qquad (2)$$

where λ_r is the wavelength of the probe beam and θ_B is the Bragg angle. Thus, changes in n, Δn and d due to analyte exposure can be indirectly measured via the η. Recently, Sabad-e-Gul et al. implemented this approach for the development of a surface relief diffraction grating (SRG)-based sensor for the detection of heavy metal ions in water [7]. The SRG-based sensor, fabricated via the holographic lithography of an acrylate photopolymer surface, which is subsequently functionalized with zeolite nanoparticles, successfully detected low concentrations of copper (Cu^{2+}), Pb^{2+} and Ca^{2+} cations. It is postulated that the obtained change in η results from the adsorption of the metal ions onto the zeolite nanoparticles on the SRG surface, thereby changing the n of the functionalizing component (i.e., the zeolite nanoparticles) and consequently changing Δn.

While this indirect measurement technique is a straightforward and fast method for sensor evaluation and characterization, this approach provides limited information on the underlying sensor operation mechanism, such as the relative contribution of simultaneous and competing changes in grating n, Δn and d to the overall measured sensor response. Moreover, due to the nature of the Bessel function in Equation (1), it is not readily possible to ascertain from a change in η alone whether φ is increasing or decreasing as a result of analyte exposure. Theoretical modelling of the processes can be conducted; however, models require assumptions and a robust model has yet to be reported. It is thus preferable to directly measure the changes in grating n and d due to analyte exposure. Such measurements will facilitate the direct study of sensor–analyte interactions, which will facilitate the enhanced understanding, design and fabrication of optical sensors.

Here, the use of ellipsometry as a characterization tool to provide further insight into optical sensor operation and zeolite–analyte interactions is presented. Ellipsometry is a highly sensitive optical technique that uses polarized light to measure the dielectric properties, such as refractive index, of a thin film or layer system [8,9]. A beam of light with a known polarization state is transmitted or reflected from the surface of the thin film, causing a change in its polarization state. The modified polarization state can be decomposed into the reflection coefficients, r_p and r_s, as derived by Fresnel, of the parallel and normal components of the electric field with respect to the plane of incidence. Ellipsometry measures the ratio of r_p and r_s (known as the ellipsometric ratio, ρ), and uses this to calculate the ellipsometric angles, Δ and ψ:

$$\rho = \frac{r_p}{r_s} = \tan \psi \times e^{i\Delta} \qquad (3)$$

The angle of incidence, θ, of the light beam is selected to be near Brewster angle of the substrate in order to maximize the difference between r_p and r_s. Following the measurement of Δ and ψ, a layer model is established for the thin film, which consists of any known optical constants (n, k) and thicknesses (d) of all individual sequential layers within the thin film. Using an iterative approach, the unknown optical constants and/or thicknesses are then varied until the best match for the measured values of Δ and ψ is obtained. For increased accuracy, as much information as possible regarding the layer model should be known in advance.

The advantages of ellipsometry are obvious; it is a non-destructive, non-contact and non-invasive characterization technique that readily achieves sub-nanometer resolution in thickness. Due to this high sensitivity, its principles have even sometimes been used as a sensor transduction mechanism [10,11].

The current study aims at developing a better understanding of the changes to the nanozeolite-doped thin film that result as a consequence of its exposure to a target analyte. In addition, consideration will be given to the advantages and limitations of both the single wavelength and spectroscopic ellipsometry apparatus for this particular study.

2. Materials and Methods

2.1. Synthesis and Characterisation of LTL-Type Zeolite Nanoparticles

The Linde Type L (LTL) zeolite suspensions were prepared in the following manner. Firstly, solution A was prepared by dissolving 2.19 g of KOH (ACS reagent, ≥85%, pellets, Sigma-Aldrich, Saint-Quentin Fallavier, France) and 0.49 Al(OH)$_3$ (reagent grade, Sigma-Aldrich) in 6.94 g of doubly distilled water at room temperature and stirred until the water was clear. Solution B was prepared by dissolving 1.09 g of KOH (ACS reagent, ≥85%, pellets, Sigma-Aldrich) and 10 g of LUDOX® SM 30 colloidal silica (30 wt.% suspension in H$_2$O, Sigma-Aldrich) in 3.47 g of doubly distilled water at room temperature and stirred until the water was clear. Afterwards, solution A was added dropwise into solution B under vigorous stirring at room temperature to achieve a non-opaque suspension, free of organic template, with the following molar composition: 5 K$_2$O:10 SiO$_2$:0.5 Al$_2$O$_3$:200 H$_2$O [12]. The as-prepared precursor mixture was aged at room temperature for 24 h prior to hydrothermal treatment at 170 °C for 18 h. Upon completing the crystallization process, the nanosized product was washed with doubly distilled water and recovered by multistep centrifugation (20,000 rpm, 40 min) until pH = 8. The final stabilized nanocrystal suspension had a concentration of approximately 1.5 wt.%.

Dynamic light scattering and X-ray diffraction measurements were carried out to determine the size of the nanoparticles and to confirm their crystalline structure, respectively. The agglomeration and average size of particles and the aggregates in the crystalline suspensions were determined using dynamic light scattering (DLS) performed with a Malvern Zetasizer Nano instrument (Malvern, UK), scattering angle of 173°, laser wavelength of 632.8 nm and output power of 3 mW). X-ray diffraction measurements were operated on a PANalytical X'Pert Pro diffractometer (Almelo, The Netherlands) using the Cu Kα monochromatized radiation (α = 1.54059 Å). A detailed crystal morphology, particle size distribution and crystallinity of the LTL zeolite were examined by field-emission scanning electron microscope (FESEM) using a TESCAN Mira (Brno, Czech Republic) operating at an accelerating voltage 30 kV, and high-resolution transmission electron microscope (HRTEM) using a Tecnai G2 30 UT (LaB6, Hillsboro, OR, USA) operated at 300 kV with 0.17 nm point resolution equipped with an EDAX EDX detector (Mahwah, NJ, USA).

2.2. Preparation of TEOS-LTL Films

As in [7], the ellipsometry measurements necessitated prolonged exposure of the zeolite films to water. In order to facilitate increased mechanical stability of the LTL-type zeolite films in solution, the LTL-type zeolites were mixed with a pre-hydrolyzed tetraethyl orthosilicate (TEOS).

The following composition was used: 24 mL TEOS, 17.5 mL ethanol, 3 mL of 0.04 M nitric acid and 12 mL of the 1.5 wt.% LTL zeolite suspension described in Section 2.1 [13]. This solution was magnetically stirred for 24 h, filtered (0.8 μm), and then spin-coated at 5000 rpm onto a Silicon wafer substrate. This procedure produced dry films with a thickness of approximately 220 nm. Reference TEOS-only (i.e., no zeolites) samples were also fabricated in the same manner. Scanning electron microscopy studies were conducted to confirm the uniform distribution of LTL-type zeolites within the TEOS-LTL film.

All samples (both TEOS-LTL and TEOS-only) were initially annealed for 3 h to ensure that the zeolite nanoparticles contained no water prior to the ellipsometric measurements. For the single wavelength ellipsometric study, the samples were annealed at 170 °C under vacuum, whereas, for the spectroscopic ellipsometric study, the samples were annealed at 170 and 320 °C in air. The second annealing temperature, 320 °C, was added in order to compensate for the difference in the annealing conditions at 170 °C (i.e., vacuum vs. air), which arose due to the difference in available equipment across the multi-national laboratories contributing to the reported work. For the data presented for each study (single wavelength and spectroscopic) identical thermal treatment conditions were used, which facilitates meaningful analysis of the data.

2.3. Ellipsometric Characterization

2.3.1. Single Wavelength Ellipsometry Experiments

A single wavelength (632.8 nm) Multiskop (Optrel GbR, Sinzing, Germany) ellipsometer was used to study the change in the optical refractive index, n, and thickness, d, of the TEOS-LTL and TEOS-only films during exposure to Cu^{2+} water solutions with concentrations of 0, 2 and 4 mM. A custom-built liquid cell with quartz windows allowed for the transmission of the 632.8 nm ellipsometer probe beam with minimal losses. A similar approach was successfully used by Pristinski et al. [14] for the study of thin film swelling in liquid environments. Each test film was placed in the cell, which was then filled with 90 mL of deionized water. The measurement was commenced as soon as the probe beam was correctly aligned with both the sample and detector. A value for the Δ and ψ parameters was captured every 10 s for 700 s. After 200 s, 10 mL of a concentrated Cu^{2+} solution was then added to the cell, ensuring not to disturb the test film. Due to dilution, the final overall concentrations in the cell were 0, 2 and 4 mM. The experiment was carried out in this fashion in order to determine if the change observed in the Δ and ψ parameters of the test films is due simply to water or if the metal solution has a separate effect.

Following data collection, a four-media model within the Optrel Elli v3.2 software was used to calculate values for the d and n of the test films as function of time. These four media from top to bottom are: water ($n = 1.33$), the test film, SiO_2 ($n = 1.4585$, $d = 3$ nm) and Si ($n = 3.8858$, $k = 0.018$ [15]). The model was additionally supplied with upper and lower limits n and d of the test film: $n = 1.35$–1.60 and $d = 140$–260 nm.

2.3.2. Spectroscopic Ellipsometry Experiments

A phase modulated spectroscopic ellipsometer (UVISEL 2, Horiba JobinYvon, Longjumeau Cedex, France) was used to characterize the change over time in the refractive index, n, the extinction coefficient, k, and the thickness, d, of the TEOS and TEOS-LTL films as a function of the incident wavelength. The phase modulated ellipsometer, which contains a Xenon light source, shows higher acquisition speed compared to, for example, null ellipsometer and rotating parts ellipsometer, because of the presence of the photoelastic modulator (PEM) with modulation frequency of 50 kHz. The modulator induces temporal change in polarization state of the light thus eliminating the need of rotating polarizer, analyser and compensator.

A specially designed cell was used in the study to enable measurements in liquid environment. In the first measurement step, the sample was placed in the cell and a measurement was taken in air. Then, without disturbing the sample, the solution of Cu^{2+} with particular concentration (0, 2 or 4 mM) was added and the measurements were taken after 90, 500 and 1000 s.

The temporal resolution of a specific scientific instrument is provided by the manufacturer and it is given for a single wavelength. However, the specific measurement time is determined by other parameters. For these measurements, scans were performed over the wavelength range 320–800 nm using a 5 nm increment. The increment was chosen based on the estimated thickness of the sample.

The time integration interval was set to 200 ms, so that an optimum signal to noise ratio was achieved, taking into account the reflectivity of the sample. Thus, the time required for a single scan at these measurement conditions was larger than 19.2 s, since some time is required to record the data at a particular wavelength and to carry out the next measurement.

For the determination of the optical constants, a four media model was implemented: silicon substrate, the studied film, a thin (1–3 nm) surface layer that contains 50% voids and water. The top layer was used for modelling the surface roughness of the studied film and its thickness is one of the parameters that was calculated. In spectroscopic ellipsometry for the determination of optical constants (n and k), the so-called dispersion models were used. They relate n and k with wavelength through different dispersion parameters that usually have physical meaning. In our case, we used the one-oscillator Lorentz model where the complex dielectric constant, ε, is described as:

$$\varepsilon = \varepsilon_\infty + \frac{(\varepsilon_s - \varepsilon_\infty)\omega_t^2}{\omega_t^2 - \omega^2 + i\Gamma_o\omega} \tag{4}$$

where ε_∞ is the high frequency dielectric constant, ε_s gives the value of the static dielectric constant at a zero frequency of light, ω is the frequency of light (in eV), ω_t (in eV) is the resonant frequency of the oscillator, whose energy corresponds to the absorption peak and Γ_o (in eV) is the broadening of the oscillator also known as damping factor. The relation between n, k and ε is:

$$\varepsilon_r = n^2 - k^2 \tag{5}$$

$$\varepsilon_i = 2nk \tag{6}$$

where ε_r and ε_i are the real and imaginary parts of ε, respectively.

3. Results and Discussion

3.1. Characterisation of the LTL Nanocrystals

The size of the LTL nanoparticles was confirmed to be 50 nm, as measured by DLS (Figure 1a). The DLS provides information on the average size of the crystals as individuals and as aggregates due to random Browning motion in the suspension during the measurements. Therefore, the particle size distribution curve covers the region 25–80 nm but centered at 50 nm. In addition, the size of individual zeolite grains according to the high-resolution transmission electron microscopy study (HRTEM) was determined to be in the range 10–20 nm (Figure 2b). It can be clearly observed that the LTL sample is fully crystalline (Figure 1b) and consists of many aggregates, which are formed from single rectangular crystalline domains with well-defined edges and crystalline fringes (Figure 2b).

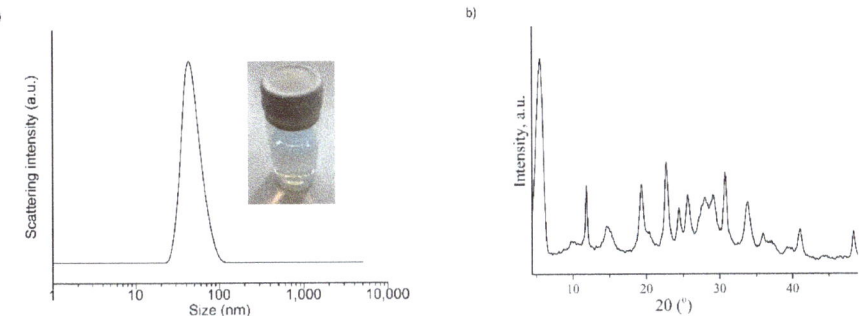

Figure 1. Dynamic light scattering (DLS) data (**a**) and the X-ray diffraction (XRD) pattern (**b**) of the Linde Type L (LTL) zeolite nanoparticles.

The SEM picture in Figure 2a displays aggregates with the size of 100 to 200 nm with the particles closely connected to each other. These agglomerates are composed of crystalline single nanocrystals of rectangular shape with prominent edges with an average size of 20 nm. The aggregated crystals in the SEM pictures are due to the drying of the LTL colloidal suspension prior to the SEM study.

Figure 2. SEM (Scale bar, M = 2000 nm) (**a**) and high-resolution transmission electron microscope (HRTEM) (Scale bar, M = 10 nm) (**b**) pictures of LTL nanosized zeolite.

3.2. Single Wavelength Ellipsometric Studies

Figure 3 shows the results from the single wavelength ellipsometric measurements carried out using both undoped and doped (i.e., TEOS-only and TEOS-LTL) samples that have been exposed to water solutions of Cu^{2+} ions with concentrations of 0, 2 and 4 mM. All samples were thermally treated in a vacuum oven at 170 °C for 3 h in order to remove any residual water from the samples as a result of ambient humidity. A significant effort was made to minimize the period required for ellipsometer beam realignment between the addition of water and the commencement of the measurement (i.e., t = 0 s in Figure 3). This period was typically 50 ± 10 s. While some swelling of the film is expected during this period, the initial absolute values for n and d were largely consistent, implying that changes occurring during this period were minimal and consistent across all samples. Specifically, for the TEOS-LTL data shown in Figure 3, the average n at t = 0 s was 1.4412 ± 0.0029; the average d was 218.83 ± 4.00 nm. For the TEOS-only samples, the average n at t = 0 s was 1.4198 ± 0.0024; the average d was 227.63 ± 7.05 nm.

3.2.1. Dynamic Studies of Film n and d due to Exposure to Water

It can be seen (Figure 3a,d) that when exposed to water, both n and d of the reference TEOS-only films were relatively constant. This could be explained by the relatively low porosity of the undoped films, which reveals that the amount of water penetrating into the films was insignificant. The change in d (Figure 3a) was within 0.12% while the change in the films n was less than 1.3×10^{-4} (Figure 3d). It is worth noting that no significant disturbance of the samples was observed as a result of the injection of water at 200 s. This confirms that there was no mechanical instability during the injection step causing changes, thus any changes observed in presence of copper in the sample can be attributed solely to the presence of the copper analyte.

The introduction of LTL zeolite nanoparticles in the film increased swelling significantly (up to 1.2% in a 220-nm thick film) and the observed change in n is 1.1×10^{-3}, revealing that the structure is more flexible (i.e., able to swell) as well as more porous and hydrophilic, which allows for the water molecules to penetrate the film.

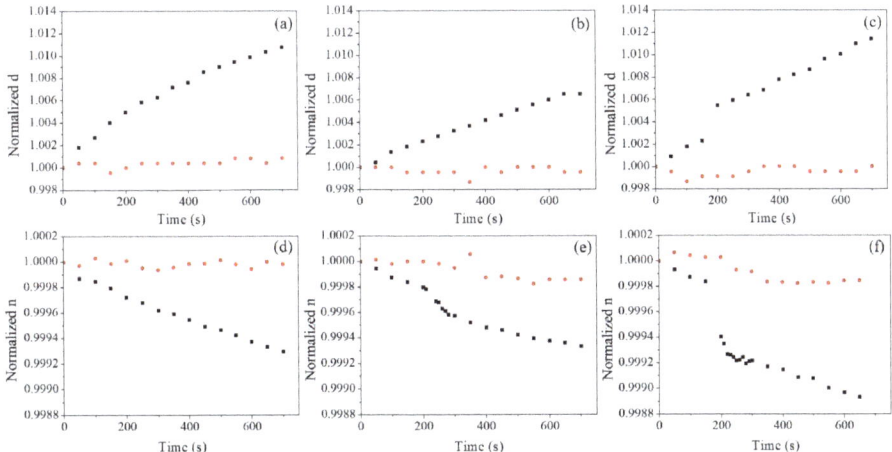

Figure 3. (Top row) normalized thickness, *d*, as a function of exposure time to 0 (**a**), 2 (**b**) and 4 (**c**) mM Cu^{2+} solution; (bottom row) normalized refractive index, *n*, as a function of exposure time to 0 (**d**), 2 (**e**) and 4 (**f**) mM Cu^{2+} solution. Black square symbol represents tetraethyl orthosilicate (TEOS)-LTL data, and red spheroidal symbol represents TEOS-only data.

3.2.2. Dynamics Studies of Film *n* and *d* due to Exposure to Copper Ions in Water

As seen from Figure 3b,c, the introduction of copper in the solution did not lead to a significant dimensional change in the TEOS-only films (within 0.13% for both 2 and 4 mM), while, in the TEOS-LTL zeolite-doped films, the swelling was similar to water solutions and within 1.2%. Nevertheless, a close examination revealed that the dynamics of the swelling is different, with a small jump at 200 s for the 4 mM concentration. One can conclude from these results that the swelling of the zeolite doped films is mainly due to the film being immersed in water and it is within 1.2% of the original thickness of the film for the first 700 s.

The examination of the dynamics of the refractive index of TEOS-LTL films under exposure to copper ions in water revealed that the rate of initial change of the refractive index after introduction of the copper ions strongly depends on their concentration (Figure 3d–f). The absolute refractive index change measured 500 s after the injection of copper ions was 7×10^{-4}, 7×10^{-4} and 11×10^{-4} for samples with a concentration of 0, 2 and 4 mM correspondingly. This can be explained by the high ion exchange capacity of the LTL zeolite nanocrystals. The higher amount of Cu cations replacing the original K cations in the LTL zeolite nanocrystals leads to a higher decrease in the absolute refractive index (Figure 3).

As *d* increases due to swelling, the average *n* of the film will decrease as the TEOS-LTL matrix molecules become more separated in space. While one would expect that water replacing air will increase *n* of the film voids, the net effect of swelling of the TEOS-LTL matrix, which has an initially higher refractive index, likely overcomes this effect, causing an overall decrease in the *n* of the film with increasing water/Cu^{2+} exposure. This effect is not observed for the TEOS-only films, which are highly rigid. Understanding these effects is important for sensor design.

3.3. Spectroscopic Ellipsometric Studies

The spectroscopic measurements were carried out with two sets of samples: the first set was thermally treated at 170 °C for 3 h i.e., identical treatment to the samples studied by single wavelength ellipsometry except that the treatment was performed in air; the second set of samples was exposed to 320 °C degrees for 1 h in air in order to suppress any dimensional changes under exposure to water and to remove water and organic residues from the pores.

The calculated extinction coefficients, k, for the films in air, water and 4 mM Cu^{2+} solution are presented in Table S1. These k values are averaged over three different samples and the deviations from the average value are in the range $(0.6–1) \times 10^{-3}$ with the exception of 3×10^{-3} for the TEOS-only (170 °C) samples. Immersion does not lead to any substantial change in k, nor is some trend observed. This demonstrates that the absorption of the films at the wavelengths used in this study is negligible.

3.3.1. Studies of Film n and d due to Exposure to Water

Initially, we studied the optical behaviour of samples in pure water. Figure 4 presents the dispersion curves of the calculated refractive index, n, of both sets of samples measured prior to and after the immersion in water, while the calculated thickness values are displayed in Table 1. It is seen from Figure 4 that the refractive index of the TEOS-LTL composites annealed at both temperatures was substantially lower compared to the TEOS-only layers. The reason is that the refractive index of the LTL zeolites (~1.37) [16] is less than the refractive index of the TEOS matrix, and thus LTL addition reduces the effective refractive index of the composite. Furthermore, the annealing at higher temperature also led to a reduction in the refractive index; this effect was most pronounced for doped samples. The most probable reason is that after 170 °C annealing in air, water was still present in the pores of zeolite particles and the TEOS matrix but was then removed after the higher temperature treatment at 320 °C.

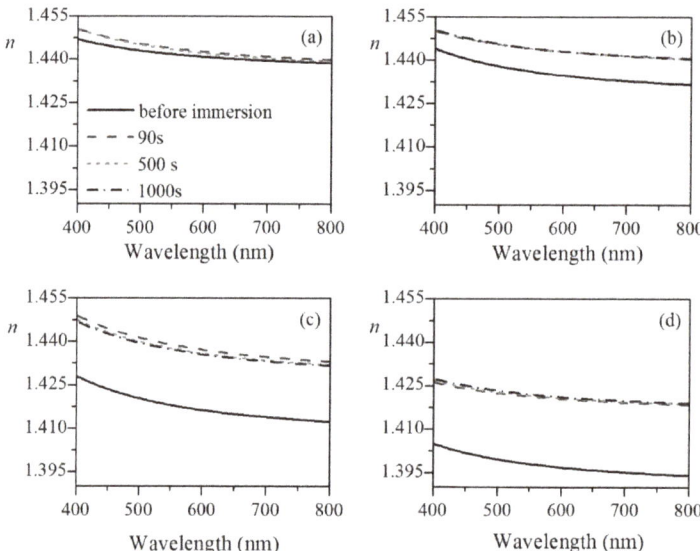

Figure 4. Dispersion curves of refractive index of TEOS-only films annealed at 170 °C (**a**) and 320 °C (**b**) and TEOS-LTL films annealed at 170 °C (**c**) and 320 °C (**d**) immersed in water for the denoted duration. The reference curve of refractive index of the layers prior to immersion is also plotted (solid black line).

Table 1. Calculated thickness values (in nm) and thickness change $\Delta d = (d_{1000s} - d_{90s})/d_{90s}$ of undoped (TEOS-only) and doped (TEOS-LTL) films annealed at 170 and 320 °C after immersion in water for 90, 500 and 1000 s.

Film Type	90 s	500 s	1000 s	Δd (%)
TEOS (170 °C)	159	159.7	159.9	0.57
TEOS (320 °C)	160.4	160.7	160.7	0.19
TEOS-LTL (170 °C)	210.6	211.1	211.3	0.33
TEOS-LTL (320 °C)	195.7	195.6	195.4	−0.15

For all samples, the immersion in water led to an increase in the refractive index; this is expected considering the presence of porosity in the layers. Water, with a higher refractive index ($n \sim 1.33$) than air ($n = 1$), penetrates the pores, thus increasing the effective refractive index. It may be expected that with increasing the immersion time this trend will continue. However, it is interesting to note that for all samples except TEOS-LTL annealed at 320 °C, n was observed to decrease with increasing time of immersion (Figure 5). The most pronounced decrease was for samples annealed at 170 °C: decrease in n from 7×10^{-4} at 500 s to 11×10^{-4} at 1000 s was observed. The reason is the swelling of the TEOS matrix with the time of immersion led to a decrease in the overall film density, thus decreasing the refractive index. From Table 1, it can be seen that the thickness of samples annealed at 170 °C increased by 0.57% and 0.33% for the undoped (TEOS-only) and doped (TEOS-LTL) films, respectively. These results suggest that the addition of LTL zeolite particles in the TEOS matrix positively influences its mechanical stability, and dimensional changes due to immersion in water are less pronounced. Additionally, the calculated thickness changes (Table 1) show that the TEOS matrix annealed at 320 °C is more rigid compared to that annealed at 170 °C; the increase in d was only 0.19% and this explains the weaker decrease in n in this case: 1×10^{-4} at 500 s to 1.4×10^{-4} at 1000 s. Furthermore, the addition of zeolites in the TEOS matrix and annealing at higher temperature contribute to an increase in the mechanical strength of TEOS-LTL samples, thus minimizing its swelling when immersed in water (thickness change is 0.15% only). In this case, an expected increase of n was observed with the time of immersion due to the constant penetration of water in the pores: 1.8×10^{-4} at 500 s to 3.6×10^{-4} at 1000 s.

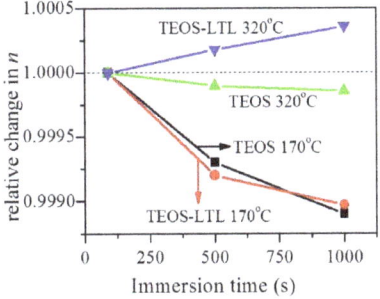

Figure 5. Relative change of refractive index (at wavelength of 800 nm) as a function of immersion time in water.

We should note that for samples annealed at 170 °C, especially these of TEOS-only, the spectroscopic ellipsometry yielded higher changes in their d and n compared to the results obtained from single-wavelength ellipsometry. The reason is that the samples are not fully identical; the first set is annealed in air, while the second set is in vacuum. The conclusion is that annealing at a low temperature (170 °C in this case) in air is not effective enough for the TEOS matrix to become sufficiently rigid, and changes related to removing organic residues and water still take place. Considering all of the above, we decided to use only samples annealed at 320 °C for further experiments.

3.3.2. Studies of Film n and d due to Exposure to Copper Ions in Water

The relative changes of n (at wavelength of 800 nm) as a function of immersion time in pure water and Cu^{2+}-containing water solution with a concentration of 4 mM for films annealed at 320 °C are presented in Figure 6. It was seen that, in both cases (water and Cu^{2+} ions), the influence of immersion was more pronounced for the TEOS-LTL samples. From the results presented above for water immersion, it has already been determined that the thickness changes in samples annealed at 320 °C are negligible and the refractive index changes are mainly due to the penetration of water inside the pores. As noted above, this is the reason for the continuous change of TEOS-LTL refractive index with immersion time, as seen in the earlier studies (Figure 5). It is also seen from Figure 6 that the changes in sample n when immersed in the copper ion solution are weaker compared to the case of water immersion. Considering that the calculated thickness changes in this case are less than 0.1%, we may conclude that the immersion of samples in Cu^{2+}-containing solution leads to a decrease in the films' refractive index for two reasons. The first reason is that the refractive index of copper solution is smaller than water [17], which leads to decrease of the effective refractive index of the films when the Cu^{2+} water solution penetrates the pores. The second reason is the decrease in the hydrophilicity of the zeolites by introducing copper; more Cu in the zeolites decreases the water content, leading to a decrease in the effective refractive index of the film. This was verified by contact angle measurement using a First Ten Angstroms (FTA200, USA) surface energy analyser (details of the measurement technique can be found in the Supplementary Material). The contact angles of the TEOS-only films and TEOS-LTL films were measured to be 62.55° and 70.50°, respectively. The increase in the contact angle value verifies the decrease in the hydrophilicity of the TEOS matrix due to the incorporation of LTL zeolites. The further increase in the contact angle of the TEOS-LTL films to 106.56° after exposure to Cu^{2+} ions in solution supports the claim that the hydrophilicity of the zeolite-doped film decreases due to the adsorption of copper (see Figure S1). Because for the TEOS-only films there is no change of hydrophilicity (no LTL zeolites), the decrease in n is smaller compared to the case of the TEOS-LTL films where both factors contribute to the decrease in n. The further increase in refractive index with time is due to the penetration of Cu^{2+} water solution inside the pores. For the TEOS-LTL samples, the difference in n of films in water and copper solution is almost the same at 500 and 1000 s—2.8×10^{-4} and 3.1×10^{-4}, respectively. The same trend is observed for the TEOS-only samples, but the changes are weaker (0.9×10^{-4} and 1×10^{-4}, respectively).

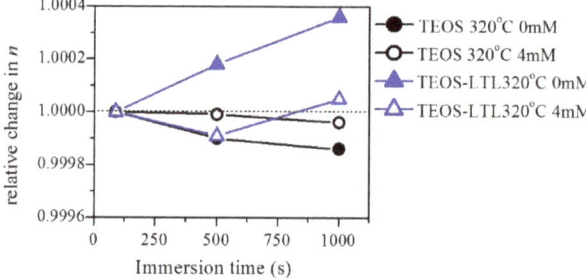

Figure 6. Relative change in refractive index, n, (at wavelength of 800 nm) as a function of immersion time in water (solid symbols) and the solution of Cu^{2+} with a concentration of 4 mM (open symbols) for films of TEOS-only (circles) and TEOS-LTL (triangles) annealed at 320 °C.

4. Conclusions

The use of ellipsometry as a characterization tool to provide further insight into optical sensor operation and zeolite–analyte interactions is presented. The suitability of both single wavelength and spectroscopic ellipsometry for the evaluation of changes in the refractive index, n, and thickness, d, of zeolite-TEOS films during exposure to copper ion solutions has been investigated. Additionally, the influence of the initial film thermal treatment conditions on the response of both doped and undoped films has been studied.

Due to the short acquisition time involved, single wavelength ellipsometry offers excellent temporal resolution and thus is suitable for the real-time study of optical changes in thin films. This approach was used to study the dynamic changes in the n and d of TEOS-only and TEOS-LTL films immersed in water and copper ion solutions. Due to the competing effects of film swelling and refractive index change due to water and copper ion adsorption, no significant difference in the absolute values of n and d were observed between the different concentrations of Cu^{2+} ions.

However, close examination reveals that the dynamics of change in n and d are different for the different Cu^{2+} concentrations, as well as between the TEOS-only and TEOS-LTL films. This highly sensitive measurement technique offers a unique insight into the sensor response mechanism over short timescales.

Spectroscopic ellipsometry was used as a second method for the study of the changes in TEOS-LTL film optical properties. While the temporal resolution of the spectroscopic system in this instance was limited by the necessity to acquire data over a larger wavelength range, the dispersion model used for data analysis in spectroscopic ellipsometry facilitates a more accurate estimation of the absolute values of the optical constants. Using this approach, detailed analysis of changes in n and d for the undoped and LTL zeolite doped TEOS films as a result of water and copper ion exposure was conducted. It was observed that the amount of film swelling can be controlled by the temperature at which the samples are treated before exposure to the analyte. The higher the temperature, the smaller the swelling of the TEOS films during immersion. Additionally, it was observed that the inclusion of LTL zeolites in the TEOS increases the rigidity and mechanical stability of the film under prolonged exposure to aqueous solutions.

Both the single wavelength and spectroscopic measurements of films pre-treated at 170 °C show a decrease in n after the sample is exposed to water or Cu^{2+}-containing water solution. Samples pre-treated at 320 °C and doped with LTL zeolites show no swelling, and n slightly increases when immersed in water and decreases after exposure to copper ions. This is explained by (i) the lower refractive index of copper water solution as compared to pure water and (ii) the decrease in the hydrophilicity of zeolites with the addition of copper.

The suitability of ellipsometry for the study of optical material and sensor–analyte interactions, including the differentiation of the influence of simultaneous and competing processes, has been demonstrated. Future work will study a wider range of zeolite–analyte combinations over a wider range of annealing temperatures and conditions.

Supplementary Materials: The following are available online at http://www.mdpi.com/2079-6412/10/4/423/s1, Figure S1: Contact angle results for uncoated and coated films: SRG-surface relief grating; SRG-TEOS-sol-gel coated surface relief grating; SRG-TEOS-LTL-zeolite-doped sol-gel coated surface relief grating; SRG-TEOS-LTL-Cu II zeolite-doped sol-gel coated surface relief grating exposed to 4 mM Cu^{2+} solution, Table S1: Extinction coefficients at wavelength of 633 nm.

Author Contributions: Conceptualization, D.C. and I.N.; methodology, D.C., T.B., V.M., S.M., S.-e.-G., A.K., C.J.B. and I.N.; formal analysis, D.C., T.B., V.M. and I.N.; resources, D.C., T.B., V.M., S.M., C.J.B. and I.N.; writing—original draft preparation, D.C., I.N. and T.B.; writing—review and editing, D.C., T.B., V.M., S.M., S.-eG., A.K., C.J.B. and I.N.; visualization, D.C., T.B. and V.M.; supervision, I.N. and C.J.B.; funding acquisition, I.N. and D.C. All authors have read and agreed to the published version of the manuscript.

Funding: This research was funded by the Arnold F Graves Postdoctoral Scholarship from Technological University Dublin, and a Flaherty Research Scholarship from the Ireland Canada University Foundation.

Acknowledgments: The authors would like to thank Oleg I. Lebedev from CRISMAT, CNRS-ENSICAEN, France, for preparation of the TEM images. The authors gratefully acknowledge the financial support provided for this work by the Arnold F. Graves Postdoctoral Scholarship from Technological University Dublin, and a Flaherty Research Scholarship from the Ireland Canada University Foundation. TB and VM acknowledge the support of the European Regional Development Fund within the Operational Programme "Science and Education for Smart Growth 2014–2020" under the Project CoE "National center of mechatronics and clean technologies" BG05M2OP001-1.001-0008-C01.

Conflicts of Interest: The authors declare no conflict of interest.

References

1. Castell, N.; Dauge, F.R.; Schneider, P.; Vogt, M.; Lerner, U.; Fishbain, B.; Broday, D.; Bartonova, A. Can commercial low-cost sensor platforms contribute to air quality monitoring and exposure estimates? *Environ. Int.* **2017**, *99*, 293–302. [CrossRef] [PubMed]

2. Yousefi, H.; Su, H.M.; Imani, S.M.; Alkhaldi, K.; Filipe, C.D.M.; Didar, T.F. Intelligent food packaging: A review of smart sensing technologies for monitoring food quality. *ACS Sens.* **2019**, *4*, 808–821. [CrossRef] [PubMed]
3. Kim, J.; Campbell, A.S.; de Ávila, B.E.; Wang, J. Wearable biosensors for healthcare monitoring. *Nat. Biotechnol.* **2019**, *37*, 389–406. [CrossRef] [PubMed]
4. Afsarimanesh, N.; Mukhopadhyay, S.C.; Kruger, M. Sensing technologies for monitoring of bone-health: A review. *Sens. Actuators A Phys.* **2018**, *274*, 165–178. [CrossRef]
5. Nascetti, A.; Mirasoli, M.; Marchegiani, E.; Zangheri, M.; Costantini, F.; Porchetta, A.; Iannascoli, L.; Lovecchio, N.; Caputo, D.; de Cesare, G.; et al. Integrated chemiluminescence-based lab-on-chip for detection of life markers in extraterrestrial environments. *Biosens. Bioelectron.* **2019**, *123*, 195–203. [CrossRef] [PubMed]
6. Raman, C.V.; Nagendra Nath, N.S. The diffraction of light by high frequency sound waves: Part 1. *Proc. Indian Acad. Sci. A* **1935**, *32*, 406–412. [CrossRef]
7. Gul, S.E.; Cody, D.; Kharchenko, A.; Martin, S.; Mintova, S.; Cassidy, J.; Naydenova, I. LTL type nanozeolites utilized in surface photonics structures for environmental sensors. *Microporous Mesoporous Mater.* **2018**, *261*, 268–274. [CrossRef]
8. Nestler, P.; Helm, C.A. Determination of refractive index and layer thickness of nm-thin films via ellipsometry. *Opt. Express* **2017**, *25*, 27077–27085. [CrossRef] [PubMed]
9. Harke, M.; Teppner, R.; Schulz, O.; Motschmann, H.; Orendi, H. Description of a single modular optical setup for ellipsometry, surface plasmons, waveguide modes, and their corresponding imaging techniques including Brewster angle microscopy. *Rev. Sci. Instrum.* **1997**, *68*, 3130–3134. [CrossRef]
10. Arwin, H. Is ellipsometry suitable for sensor applications? *Sens. Actuators A Phys.* **2001**, *92*, 43–51. [CrossRef]
11. Moirangthem, R.S.; Chang, Y.C.; Hsu, S.H.; Wei, P.K. Surface plasmon resonance ellipsometry based sensor for studying biomolecular interaction. *Biosens. Bioelectron.* **2010**, *25*, 2633–2638. [CrossRef] [PubMed]
12. Hölzl, M.; Mintova, S.; Bein, T. Colloidal LTL zeolite synthesized under microwave irradiation. *Surf. Sci. Catal.* **2005**, *158*, 11–18.
13. Jaiswal, S.; McHale, P.; Duffy, B. Preparation and rapid analysis of antibacterial silver, copper and zinc doped sol-gel surfaces. *Coll. Surf. B Biointerf.* **2012**, *94*, 170–176. [CrossRef] [PubMed]
14. Pristinski, D.; Kozlovskaya, V.; Sukhishvili, S.A. Determination of film thickness and refractive index in one measurement of phase-modulated ellipsometry. *JOSA A* **2006**, *23*, 2639–2644. [CrossRef] [PubMed]
15. Adachi, S. Model dielectric constants of Si and Ge. *Phys. Rev. B* **1988**, *38*, 12966. [CrossRef] [PubMed]
16. Babeva, T.; Awala, H.; Vasileva, M.; Fallah, J.; El Lazarova, K.; Thomas, S.; Mintova, S. Zeolite films as building blocks for antireflective coatings and vapor responsive Bragg stacks. *Dalton Trans.* **2014**, *43*, 8868–8876. [CrossRef] [PubMed]
17. Chah, S.; Yi, J.; Zare, R.N. Surface plasmon resonance analysis of aqueous mercuric ions. *Sens. Actuators B* **2004**, *99*, 216–222. [CrossRef]

© 2020 by the authors. Licensee MDPI, Basel, Switzerland. This article is an open access article distributed under the terms and conditions of the Creative Commons Attribution (CC BY) license (http://creativecommons.org/licenses/by/4.0/).

Article

Amphiphilic Poly(vinyl Alcohol) Copolymers Designed for Optical Sensor Applications—Synthesis and Properties

Katerina Lazarova [1,*], Silvia Bozhilova [2], Christo Novakov [2], Darinka Christova [2] and Tsvetanka Babeva [1,*]

[1] Institute of Optical Materials and Technologies "Akad. J. Malinowski", Bulgarian Academy of Sciences, Akad. G. Bonchev str., bl. 109, 1113 Sofia, Bulgaria
[2] Institute of Polymers, Bulgarian Academy of Sciences, Akad. G. Bonchev Str., bl. 103-A, 1113 Sofia, Bulgaria; s.bozhilova@polymer.bas.bg (S.B.); hnovakov@polymer.bas.bg (C.N.); dchristo@polymer.bas.bg (D.C.)
* Correspondence: klazarova@iomt.bas.bg (K.L.); babeva@iomt.bas.bg (T.B.); Tel.: +359-2-979-3526 (K.L.)

Received: 22 April 2020; Accepted: 7 May 2020; Published: 9 May 2020

Abstract: A possible approach for enhancement of Poly(vinyl alcohol) (PVA) humidity-sensing performance using hydrophobically modified PVA copolymers is studied. Series of poly(vinylalcohol-*co*-vinylacetal)s (PVA–Ac) of acetal content in the range 18%–28% are synthesized by partial acetalization of hydroxyl groups of PVA with acetaldehyde and thin films are deposited by spin-coating using silicon substrates and glass substrates covered with Au–Pd thin film with thickness of 30 nm. Sensing properties are probed through reflectance measurements at relative humidity (RH) in the range 5%–95% RH. The influence of film thickness, post-deposition annealing temperature, and substrate type/configuration on hysteresis, sensitivity, and accuracy/resolution of humidity sensing is studied for partially acetalized PVA copolymer films, and comparison with neat PVA is made. Enhancement of sensing behavior through preparation of polymer–silica hybrids is demonstrated. The possibility of color sensing is discussed.

Keywords: poly(vinyl alcohol) copolymers; thin films; humidity sensing; optical sensors

1. Introduction

Poly(vinyl alcohol) (PVA) is a hydrophilic and very water-soluble polymer due to pendant hydroxyl groups, which are mainly responsible for its reactivity and crystallinity [1]. Due to its outstanding mechanical and film-forming properties PVA is used in a variety of areas such as membranes, adhesives, coatings, etc. [2]. Because PVA can absorb and desorb water quickly, an increasing interest in the research of PVA-based humidity sensors has been observed recently and PVA is implemented as a humidity-sensitive medium in various sensor types [3–8].

However, development of PVA-based optical humidity sensors with high sensitivity, wide dynamic range and linearity, stability, and low hysteresis is still a challenge. This is probably due to the highly water-soluble nature of PVA that limits its stable sensing properties in a form of nanometer-thick polymer film. One possible approach to overcome these drawbacks is to use different composites consisting of PVA as a matrix. Mixture of PVA and graphene quantum dots (GQDs) [9–11] and crosslinked PVA/functionalized graphene oxide nanocomposite films [12] were used for humidity-sensing using optical fiber technology [9–11] and attenuated reflectance measurements [12]. Composites consisting of PVA and nanosilica particles were used for humidity sensing through crystal microbalance [13] or for depositing sensitive opal structures [14]. Polyaniline/poly(vinyl alcohol) composites [15] and silver–polyaniline/polyvinyl alcohol composites [16] were used as sensitive media for acoustic wave-impedance humidity sensors and resistive humidity sensors, respectively. Thick PVA substrates (around 80 microns) doped with silver nanoparticles were applied for humidity sensing through transmittance measurements [17].

Another possible approach to enhance PVA sensing performance is to use hydrophobically modified PVA copolymers such as poly(vinyl acetal)s [18]. In general, poly(vinyl acetal)s are class of polymers obtained by reaction of PVA with aldehydes, especially formaldehyde, acetaldehyde, and butyraldehyde finding advanced application as structural adhesives in the aircraft industry, as the interlayer in automotive safety glass, etc. [19]. The contents of unreacted hydroxyl groups along with the acetal rings and the molecular weight determine the polymer properties. Hydrophilicity of partially acetalized PVA is reduced while maintaining its inherent response to humidity.

Our previous studies [20,21] have shown that hydrophobically modified PVA copolymers, namely poly(vinylalcohol-co-vinylacetal)s (PVA–Ac) of acetal content in the range 18%–28% prepared as a single films on opaque substrate are suitable for optical sensing of humidity. In this work, the influence of film thickness, post-deposition annealing and substrate type/configuration on humidity-sensing properties is studied for partially acetalized PVA films and comparison with neat PVA is made. The possibility to improve the reaction toward humidity of the polymer thin films via doping with SiO_2 particles is explored and polymer–silica hybrids are obtained. Enhancement of sensing behavior through this approach is demonstrated and discussed.

2. Materials and Methods

2.1. Synthesis of PVA Copolymers

Series of hydrophobically modified PVA copolymers were synthesized by partial acetalization of hydroxyl groups of PVA (average polymerization degree 1600) with acetaldehyde following the procedure described elsewhere [18]. The copolymer composition of obtained PVA–Ac, namely the content of acetal groups, was estimated by using Nuclear Magnetic Resonance (NMR) spectroscopy. ^1H NMR spectra were taken on a Bruker Avance DRX 250 spectrometer (Bruker Corporation, Billerica, MA, USA) in DMSO-d_6 as solvent. The extent of hydrophobic modification was evaluated by means of UV-VIS spectroscopy. Transmittance of copolymer aqueous solutions was studied at wavelength of 500 nm at concentration of 5 g/L as a function of temperature. Cloud points (T_{CP}) of copolymers solutions were determined from the measured transmittance-vs-temperature curves as the temperature at transmittance level of 50%.

Copolymer solutions of 1 wt.% concentration in mixed water–methanol solvent (20:80 volume ratio) were prepared for thin film deposition process. To obtain hybrid polymer–silica thin films, SiO_2 particles were in situ generated in copolymer solutions via the sol-gel method [22,23]. Calculated amount of the precursor tetraethyl orthosilicate (TEOS) was added to the copolymer solution under stirring and mixture was acidified with 1 M HCl to pH 2. The reaction mixture was homogenized by ultra-sonication for 30 min and then TEOS hydrolysis was continued at vigorous stirring on a magnetic stirrer at room temperature for 24 h. The obtained copolymer solution doped with SiO_2 particles was used for thin films deposition without further treatment. The condensation of the silica was completed during the annealing of the deposited thin hybrid films.

2.2. Deposition of Thin Films

Water–methanol solutions in a volume ratio of 20:80 and concentrations of 1 and 2 wt.% were used for deposition of acetal modified PVA films. Thin polymer films were deposited by spin-coating method at a rotation speed of 4000 rpm and time of 60 s using 0.250 mL of the solution. After deposition, the films were annealed in air for 30 min at 60 and 180 °C. For comparison, films of neat PVA were also prepared by depositing 2 and 5 wt.% water solution of PVA to achieve the same film thicknesses as in the case of modified films. Silicon wafers and Au–Pd covered optical glass plates were used as substrates. The Au–Pd sublayers with Au:Pd ratio of 80:20 and thickness of 30 nm were deposited on glass substrates by cathode sputtering of gold/palladium target (Quorum Technologies, Lewes, UK) for

60 s under vacuum 4×10^{-2} mbar using Mini Sputter Coater SC7620 system (Quorum Technologies, Lewes, UK).

Polymer thin films composites (polymer doped with SiO_2 particles) were spin-coated on silicon wafers (0.250 mL solution at concentration of 1 wt.%, 4000 rpm, 60 s) and post-annealed at 60 °C for 30 min.

2.3. Characterization of Thin Films

Optical constants (refractive index n and extinction coefficient k) and thickness of the films d were calculated simultaneously using previously developed two-stages nonlinear curve fitting method using measured reflectance spectra with UV-VIS-NIR (ultraviolet-visible-near infrared) spectrophotometer (Cary 5E, Varian, Australia) [24]. The sensing properties of the films were studied through recording reflectance spectra at different values of relative humidity (RH) in the range from 5% to 95% RH. The sample was placed in a quartz cell inside the spectrophotometer and the humidity decreased from ambient to 5% *RH* by purging dry argon in the cell. Then the recording of reflectance (or transmittance) value as a function of humidity was started. The continuous increase of humidity from 5% to 95% RH was achieved by bubbling argon through distilled water kept at 60 °C. In these experiments the reflectance/transmittance was measured at fixed wavelength that is preliminary chosen as the wavelength of the highest humidity responses. To determine this wavelength for each thin film (λ_{max}), along with optical constants and thickness (and its change), the reflectance spectra (320–800 nm) of the samples were measured at humidity of 5% and 95% RH in another set of humidity experiments and optical constants and thickness were determined.

To quantify and compare studied samples, three parameters were used. The sensitivity of the sensors, S, was calculated according to the following equation:

$$S = \frac{\Delta R}{RH_2 - RH_1}, \qquad (1)$$

where ΔR (or ΔT, if transmittance T is measured) is the change of film's reflectance (or transmittance) in % for humidity variation from RH_1 to RH_2. Accuracy/resolution (ΔRH) of detection depends on the sensitivity and measurement accuracy in the signal and was calculated from:

$$\Delta RH = \frac{errR\ (\%)}{S\ (\%)}, \qquad (2)$$

where $errR = 0.3\%$ (or $errT = 0.1\%$, if T is measured) is the experimental error (accuracy) of R or T and S is the sensitivity, calculated by Equation (1).

Sometimes it is possible unwanted hysteresis to occur that is expressed in different values of R (or T) measured at the same values of humidity depending of the history of humidity, i.e., depending whether humidity increases or decreases. The percentage of hysteresis, H was determined through:

$$H(\%) = \frac{max|R_{up} - R_{down}|}{\Delta R_{max}} \cdot \frac{\Delta RH_{hyst}}{\Delta RH} \cdot 100, \qquad (3)$$

where R_{up} and R_{down} are reflectance (or transmittance) values measured for increasing and decreasing humidity, respectively, ΔR_{max} is the reflectance (or transmittance) change in the whole range ΔRH of measured humidity and ΔRH_{hyst} is the humidity range where hysteresis is observed.

It is obvious from Equations (1)–(3) that the goal is to obtain the highest sensitivity and accuracy of detection and the lowest percentage of hysteresis.

3. Results

3.1. Characterization of Synthesized Polymers

Four PVA–Ac copolymers of different composition were synthesized varying PVA-to-acetaldehyde molar ratio. The reaction scheme and chemical structure of the obtained PVA copolymers are illustrated in Figure 1a. The copolymer composition and aqueous solution properties were studied by NMR and UV-VIS spectroscopy, respectively. The results are summarized in Table 1.

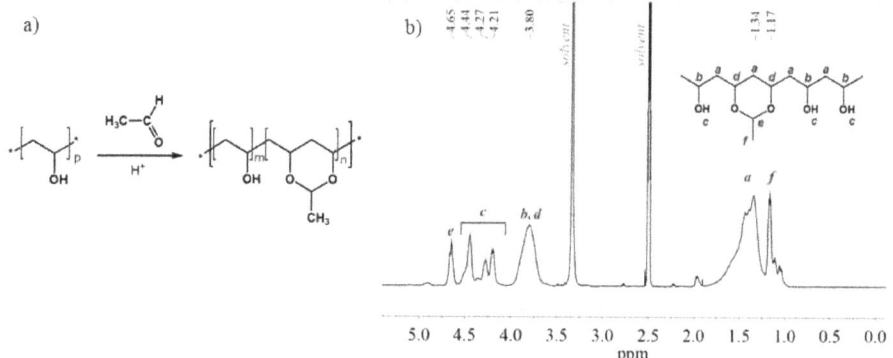

Figure 1. (a) Schematic presentation of acetalization reaction of PVA; (b) ^{1}H NMR spectrum of copolymer AC18 (solvent DMSO-d$_6$).

Table 1. Acetal content and cloud point (T_{CP}) of modified PVA copolymers.

Sample	Acetal Content, % (NMR)	T_{CP} *, °C (UV-VIS)
PVA	0	_**
Ac18	18	47
Ac19	19	40
Ac24	24	30
Ac28	28	27

* As measured for 5 g/L aqueous copolymer solution; ** No T_{CP} detected up to 90 °C.

Typical proton NMR spectrum of PVA–Ac is shown in Figure 1b. Copolymer composition expressed as a content of acetal groups was calculated by comparing the area of the peak assigned to the methine protons from the acetal group (e) to those assigned to the methine protons from the PVA main chain (b+d).

The synthesized PVA–Ac copolymers, although very water soluble at room temperature, undergo phase transition when increasing the temperature of the aqueous solutions and turn water insoluble. This is due to the introduced fractions of hydrophobic acetal groups and reflects in the reduced hydrogen bonding between copolymers and water molecules as compared to pure PVA. To evaluate the hydrophilic–hydrophobic balance of the copolymers, expected to influence the humidity-sensing properties of the corresponding thin films, T_{CP} in dilute aqueous solutions were measured. Clouding curves of the copolymer aqueous solutions were registered, and T_{CP} were estimated at 50% transmittance. As seen in Table 1, the higher the acetal content, the lower the T_{CP}.

3.2. Optimization of Thickness and Post-Deposition Annealing

When polymer films are exposed to humidity they change their thicknesses and refractive indices. This results in change of the measured reflectance or transmittance spectra. We have recently shown that the dimensional change in polymer films with nanometer thickness in the range 100–400 nm depends on the initial thickness and increases with increasing thickness [25]. Figure 2a presents the change in thickness for studied samples (80 and 200 nm) at their exposure from low to high humidity. As expected, the degree of swelling of thicker films (200 nm) is substantially higher as compared to thinner films especially for 19% modified PVA films where the relative increase of thickness ($\Delta d/d$) is 120%. Furthermore, the post-deposition thermal treatment of samples at higher temperature (180 °C) does not lead to an improvement of swelling, as we have already shown in [21]. On the contrary, the dimensional changes of films pre-annealed at 180 °C are smaller as compared to those treated at 60 °C, especially for the neat PVA films which degree of swelling is almost 7 times lowered.

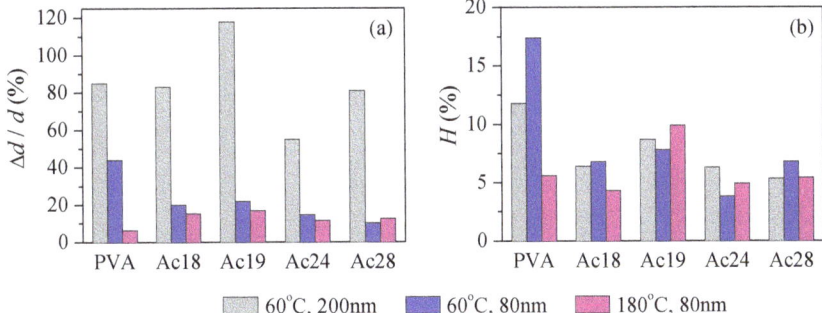

Figure 2. (a) Thickness change upon humidity exposure ranging from 5% to 95% RH of polymer films with thickness of 80 and 200 nm and of different acetal content pre-annealed at 60 and 180 °C; (b) Percentage of hysteresis, H calculated with eq. 3 for polymer films (80 and 200 nm) and of different acetal content pre-annealed at 60 and 180 °C.

It is well known that the hysteresis, H, is another very important parameter that determines the suitability of the material for sensor applications. The existence of H means measuring of different signal (reflectance or transmittance in our case) for the same humidity values depending whether humidity increases or decreases. It is obvious that H is unwanted parameter and the main goal is to keep its value as low as possible.

The hysteresis values H of all samples studied is summarized in Figure 2b and Table 2. A substantial decrease of hysteresis due to annealing at 180 °C is observed for neat PVA films. The smallest H-values (4.3% and 3.8%) are achieved for PVA-modified samples (80 nm) with acetal content of 18% and 24%, pre-annealed at 180 and 60 °C, respectively.

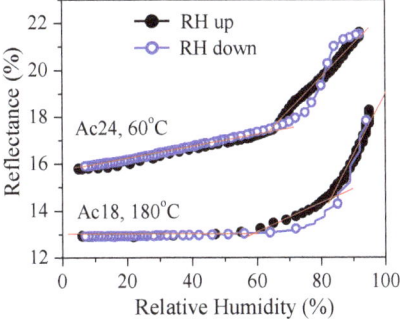

Figure 3. Reflectance versus relative humidity curves for films with 24% and 18% acetal content (80 nm thick), pre-annealed at 60 and 180 °C, respectively, measured for increasing (solid black symbols) and decreasing (open blue symbols) humidity.

Table 2. Post-annealing temperature (T_{post}), wavelength at which R (T) measurements were conducted (λ_{max}), percentage of hysteresis (H), dynamic range, sensitivity, and accuracy for studied samples.

Sample	T_{post}(°C)	λ_{max} (nm)	H (%)	Dynamic Range (% RH)	Sensitivity (%/% RH)	Accuracy (% RH)
PVA	60	597	17.4	<30	<0.01	>30
PVA	180	402	5.6	<75	0.013	23
Ac18	60	592	6.8	<45	0.010	30
Ac18	180	400	4.3	>60	0.07 (60%–84% RH) 0.3 (>84% RH)	4 1
Ac19	60	400	7.8	<40	<0.01	>30
Ac19	180	400	9.9	<60	<0.01	>30
Ac24	60	408	3.8	full	0.03 (5%–65% RH) 0.14 (> 65% RH)	10 2
Ac24	180	406	4.9	full	0.01 (5%–60% RH) 0.02 (60%–77% RH) 0.08 (>77% RH)	30 15 4
Ac28	60	598	6.8	<60	< 0.01	>30
Ac28	180	600	5.4	full	0.01 (5%–70% RH) 0.05(>70% RH)	30 6
Ac24 (T%)	60	460	3.6	full	0.03 (5%–70% RH) 0.14(>70% RH)	3 0.7
Ac24p1	60	424	1.7	full	0.03	10
Ac24p2	60	482	3.9	full	0.02	15

One can conclude that the most suitable samples are Ac24 and Ac18 annealed at 60 and 180 °C, respectively. The dependence of reflectance on relative humidity in the range 5%–95% RH (R-vs-RH curves) for both samples of the smallest H-values are presented in Figure 3.

It is seen that reflectance for Ac18 is almost the same in wide humidity range (5%–60% RH) and starts to increase exponentially at RH > 60%. On the contrary, for sample Ac24 two linear dependences of R-vs-RH plot with different slopes are well distinguished. The sensitivity is 0.03 at RH = 5%–65% and increases to 0.14 at RH > 65%. Considering that the measurement error in reflectance is 0.3% the accuracy of humidity measurement is 10% and 2% RH, respectively (Table 2). We should note that although the swelling is the strongest for thick films (200 nm) they are not suitable for sensing because exhibit high values of hysteresis (Figure 2b). Additional measurements of reflectance as a function of relative humidity (not shown) demonstrated that thicker samples have narrow dynamic range. Generally, for thicker samples R changes only for humidity higher than 80% RH. For sample Ac19 the case is even worse because there is unambiguity—one and the same reflectance values are measured for different humidity. The reason is the periodicity of the dependence of R on d. When the change of thickness due to humidity is higher than the period of R-vs-d dependence then a periodicity in R-vs-RH curve could be observed. Usually this unambiguous behavior appears for thicker films (d higher than 150 nm) where swelling is stronger as compared to thinner ones [25].

Considering all results presented above we concluded that the most appropriate sample for our purposes is Ac24 (24% acetal content PVA) with approximate thickness of 80 nm pre-annealed at 60 °C. Further efforts are concentrated on optimization of sensing properties of this material using two approaches: (i) humidity sensing through transmittance measurements; and (ii) doping with SiO_2 particles.

3.3. Humidity Sensing Using Transmittance Measurements

It is well known that in general case measuring the transmittance is easier, more accurate, and less expensive than measuring the reflectance. Therefore, it will be more advantageous to use transmittance measurements as optical read-out for detecting humidity. To perform transmittance measurements, the sensitive medium should be deposited on transparent substrate. Usually these are glass or plastic with approximate refractive index in the range 1.4–1.5 that is very close to refractive index of the polymers used for detection. Thus, the small optical contrast will lead to low sensitivity of detection, because it will be difficult to distinguish the thin film from the substrate because of the match of their refractive indices. We have already shown that quarter-wavelength multilayers stacks (Bragg stacks) and glass covered with thin semitransparent metal overlayer are suitable transparent substrates for optical detecting of humidity in transmittance mode [25,26]. When planar Bragg stacks are used for substrates, the sensitivity of detection increases with thickness of the sensitive medium deposited on top and it is the highest for films thicker than 250 nm [25]. However, as already mentioned an ambiguity exists when films with thicknesses higher than 100–150 nm are used as sensitive media. Therefore, in this study we use thin film with thickness of 80 nm deposited on Au–Pd covered glass substrate. The thickness of the metal overlayer is selected to be 30 nm thus guaranteeing transmittance to be around 50%.

The transmittance of Ac24 thin film with thickness of 80 nm, deposited on glass covered with Au–Pd overlayer with thickness of 30 nm as a function of relative humidity is shown in Figure 4. The observed percentage of hysteresis of 3.6% is very close to the value obtained when reflectance as a function of humidity is used (3.8%) (Figure 2b). Furthermore, similarly to the case of reflectance measurements (Figure 2b), two well-distinguished linear parts of the T-vs-RH curve are observed. The calculated sensitivities are comparable to the case of silicon substrate: 0.03 in the range 5%–70% RH and 0.14 for $RH > 70\%$. However, because of the higher accuracy in transmittance measurements (errT is 0.1% as compared to errR = 0.3% when R is measured) the accuracy of sensing ΔRH is 3 times higher (Equation (2)). Thus, using the configuration polymer film/metal layer/glass, less than 1% RH could be distinguished in the range of high humidity ($RH > 70\%$) and 3% for $RH < 70\%$ (Table 2).

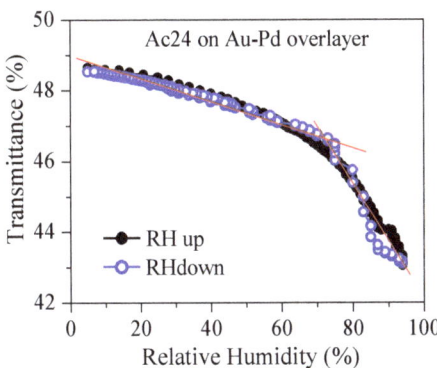

Figure 4. Transmittance versus relative humidity curve for PVA film (80 nm thick) of 24% acetal content, deposited on Au–Pd (30nm) covered glass substrate, pre-annealed at 60 °C measured for increasing (solid black symbols) and decreasing (open blue symbols) humidity.

3.4. Doping with SiO_2 Particles

Incorporation of silica nanoparticles may further enhance the sensing properties of studied PVA–Ac copolymer thin films. The silanol groups on the particles' surface can develop intermolecular bonds with PVA hydroxyl groups and may assist improving the sensing performance and reducing hysteresis. This approach was demonstrated by implementing in situ sol-gel reaction of TEOS in Ac24 copolymer solutions prior to the thin film deposition.

The influence of SiO_2 particles doping on humidity thickness change and percentage of hysteresis is illustrated in Figure 5a. The comparison between undoped (Ac24) and differently doped samples (Ac24p1, 20% and Ac24p2, 50%) shows that the swelling ($\Delta d/d$) due to humidity exposure decreases with doping from 14.5% for undoped sample to 6.6% and 4.2% for 20% and 50% doped samples, respectively. The possible reason is the rigidity of the films that increases when SiO_2 particles are incorporated in the polymer matrix. However, the doping has a positive effect on percentage of hysteresis: H decreases more than twice for 20% doped sample (Ac24p1): from 3.8% (undoped film) to 1.7% (20% doped film). In this case, the increased rigidity of the doped film contributes positively because it prevents fast shrinking of polymer films during humidity desorption in high range (RH > 70%) where the hysteresis is commonly observed.

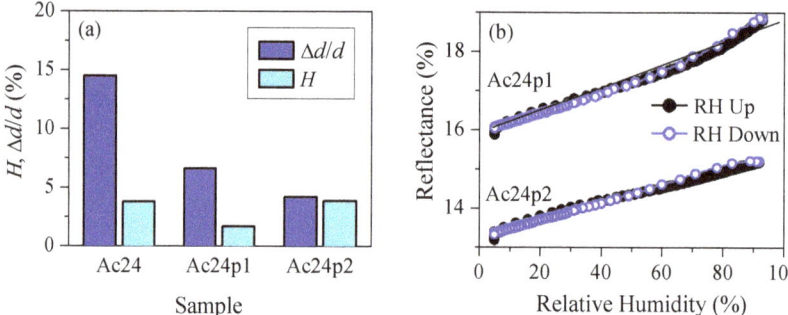

Figure 5. (a) Percentage of hysteresis, H (Equation (3)) and relative change of film thickness $\Delta d/d$ due to humidity exposure from 5% to 95% RH for polymer films (about 80 nm in thickness) with 24% acetal content (Ac24) doped with 20% (Ac24p1) and 50% (Ac24p2) SiO_2 particles; (b) Reflectance versus relative humidity curve for films with 24% acetal content doped with 20% and 50% SiO_2 particles measured for increasing (solid black symbols) and decreasing (open blue symbols) humidity.

The R-vs-RH curves for the doped films are plotted in Figure 5b. A very good linear dependence of measured reflectivity is observed in the whole RH range for both samples, mostly pronounced for heavily doped one (Ac24p2) sample. This means that the acetal modified PVA films doped with appropriate amount of SiO_2 particles are very suitable for optical sensing of humidity offering linearity in the entire humidity range.

Results summary is presented in Table 2. From all studied samples the smallest hysteresis is observed for Ac24 polymer film doped with 20% SiO_2 particles. The sample has full dynamic range and linear dependence of the measured signal as a function of relative humidity. The accuracy of sensing is 10% RH but could be substantially increased if transmittance measurements are implemented. More studies are underway to optimize the sensing behavior of doped polymer samples.

It is seen from Table 2 that another good option for optical sensing of humidity is Ac24 polymer film (PVA with 24% acetal content) deposited on glass substrate with Au–Pd overlayer (30 nm). At RH > 70% the accuracy is less than 1% RH, the dynamic range is wide, and the hysteresis is acceptable.

3.5. Color Sensing of Humidity

When transparent film is deposited on absorbing substrate (silicon wafer or metal overlayer in our case) it exhibits a certain color which depends on film's optical thickness. When optical thickness changes due to adsorption of water vapors (as is in our case) from the environment, the reflectance spectrum of the film changes and the color alters. Optical sensing based on perceptual color change in response to analyte of interest offers simplicity and is preferred in color sensing of vapors. So the next step of our investigation was to study the possibility of color sensing of humidity, i.e., monitoring of color at different humidity levels for Ac24 thin film on both types of substrates (Si wafer and glass covered with Au–Pd thin film) post-annealed at 60 °C. Figure 6 presents the calculated color coordinates (CIE X and CIE Y) of PVA films with 24% acetal content deposited on selected substrates for low (5% RH) as well as high humidity (95% RH). In the case of film on Si substrate, reflectance spectra are used for calculation, while for polymer film deposited on Au–Pd/glass, transmittance spectra at 5% and 95% RH are used. It is seen from Figure 6 that a substantial change of color takes place in the first case: the two points, for 5% and 95% RH, are well separated in the color space. On the contrary, when transparent substrate is used the change of the color due to humidity is not so distinct: the two points, associated with the sample colors at low and high humidity, respectively, almost overlap.

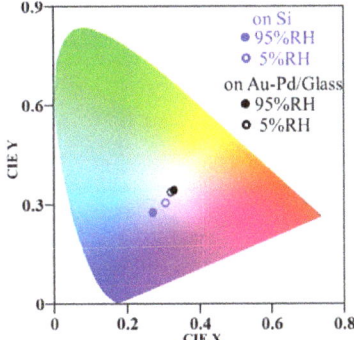

Figure 6. Calculated CIE coordinates for Ac24 thin film deposited on opaque (Si) and transparent (glass covered with Au–Pd thin film) substrates exposed to 5% and 95% RH.

4. Conclusions

The successful humidity-sensing application of thin films of hydrophobically modified PVA copolymers, namely poly(vinylalcohol-*co*-vinylacetal)s (PVA–Ac), of acetal content in the range of 18%–28% is demonstrated. A noticeable decrease of hysteresis, increase of sensitivity, and widening of dynamic range for modified films as compared to the neat PVA films are observed. For optimization of sensor performance, post-deposition annealing at 60 and 180 °C is used and two different film thicknesses are used (80 and 200 nm). The best sensor characteristics are obtained for films modified with acetal content around 24% with thickness of 80 nm and post-deposition annealing temperature of 60 °C. For relative humidity higher than 70% an accuracy of 0.7% RH is achieved. Both post-deposition annealing at 180 °C and higher film thickness (200 nm) leads to deterioration of sensing operation of the films.

It was demonstrated that both types of substrate used, silicon substrate and glass with thin (30 nm) metal (Au–Pd) overlayer, are suitable for humidity sensing. The first one is preferable if color sensing of humidity is considered, while the second one enables transmittance measurements thus offering more technological convenience and higher accuracy/resolution of measurements. For further decrease of hysteresis, a doping of PVA–Ac (24%) with SiO_2 particles (20%) is used. The thin film samples have full dynamic range and linear dependence of the measured signal in the entire humidity range. Humidity-sensitive films have thickness values around 80 nm that guarantees fast sensing.

Author Contributions: Conceptualization, T.B., D.C., and C.N.; methodology, T.B., D.C., and K.L.; software, T.B. and K.L.; validation, T.B., D.C., C.N., and K.L.; formal analysis, T.B., D.C., and K.L.; investigation, K.L. and S.B.; resources, D.C. and T.B.; data curation, K.L., D.C., and T.B.; writing—original draft preparation, T.B. and D.C.; writing—review and editing, T.B., D.C., and K.L.; visualization, T.B. and D.C.; supervision, T.B. and D.C.; project administration, T.B. and D.C.; All authors have read and agreed to the published version of the manuscript.

Funding: This research was funded by Bulgarian National Science Fund, Grant No. DN08-15/14.12.2016.

Acknowledgments: K.L. and S.B. acknowledge the National Scientific Program for young scientists and postdoctoral fellows, funded by Bulgarian Ministry of Education and Science (PMC № 271/2019). This work was partially supported by the European Regional Development Fund within the Operational Programme "Science and Education for Smart Growth 2014–2020" under the Project CoE "National center of mechatronics and clean technologies" BG05M2OP001-1.001-0008-C01.

Conflicts of Interest: The authors declare no conflict of interest. The funders had no role in the design of the study; in the collection, analyses, or interpretation of data; in the writing of the manuscript, or in the decision to publish the results.

References

1. Finch, C.A. (Ed.) *Polyvinyl Alcohol: Developments*, 2nd ed.; John Wiley & Sons Ltd.: Chichester, UK, 1992; ISBN 978-0-471-99850-1.
2. Farahani, H.; Wagiran, R.; Hamidon, M.N. Humidity Sensors Principle, Mechanism, and Fabrication Technologies: A Comprehensive Review. *Sensors* **2014**, *14*, 7881–7939. [CrossRef]
3. Zhao, C.; Yuan, Q.; Fang, L.; Gan, X.; Zhao, J. High-performance humidity sensor based on a polyvinyl alcohol-coated photonic crystal cavity. *Opt. Lett.* **2016**, *41*, 5515–5518. [CrossRef] [PubMed]
4. Amin, M.D.; Karmakar, N.; Winther-Jensen, B. Polyvinyl-alcohol (PVA)-based RF humidity Sensor in microwave frequency. *Prog. Electromagn. Res. B* **2013**, *54*, 149–166. [CrossRef]
5. Jang, J.H.; Han, J.I. Cylindrical relative humidity sensor based on poly-vinyl alcohol (PVA) for wearable computing devices with enhanced sensitivity. *Sensors Actuators A Phys.* **2017**, *261*, 268–273. [CrossRef]
6. Ogura, K.; Patil, R.C.; Shiigi, H.; Tonosaki, T.; Nakayama, M. Response of Protonic Acid-Doped Poly(o-Anisidine)/Poly(Vinyl Alcohol) Composites to Relative Humidity and Role of Dopant Anions. *J. Polym. Sci. Part A Polym. Chem.* **2000**, *38*, 4343–4352. [CrossRef]
7. Alam, S.; Islam, T.; Mittal, U. A Sensitive Inexpensive SAW Sensor for Wide Range Humidity Measurement. *IEEE Sens. J.* **2020**, *20*, 546–551. [CrossRef]
8. Lin, H.; Liu, F.; Dai, Y.; Mumtaz, F. Relative humidity sensor based on FISM-SMS fiber structure coated with PVA film. *Optik* **2020**, *207*, 164320. [CrossRef]
9. Tong, R.-J.; Zhao, Y.; Zheng, H.-K.; Xia, F. Simultaneous measurement of temperature and relative humidity by compact Mach-Zehnder interferometer and Fabry-Perot interferometer. *Measurement* **2020**, *155*, 107499. [CrossRef]
10. Zhao, Y.; Tong, R.-J.; Chen, M.-Q.; Xia, F. Relative humidity sensor based on hollow core fiber filled with GQDs-PVA. *Sens. Actuators B Chem.* **2019**, *284*, 96–102. [CrossRef]
11. Tong, R.-J.; Zhao, Y.; Chen, M.-Q.; Peng, Y. Multimode interferometer based on no-core fiber with GQDs-PVA composite coating for relative humidity sensing. *Opt. Fiber Technol.* **2019**, *48*, 242–247. [CrossRef]
12. Mallakpour, S.; Abdolmaleki, A.; Khalesi, Z. Fabrication and physicochemical features study of crosslinked PVA/FGO nanocomposite films. *Polym. Bull.* **2018**, *75*, 1473–1486. [CrossRef]
13. Zheng, X.; Fan, R.; Li, C.; Yang, X.; Li, H.; Lin, J.; Zhou, X.; Lv, R. A fast-response and highly linear humidity sensor based on quartz crystal microbalance. *Sens. Actuators B Chem.* **2019**, *283*, 659–665. [CrossRef]
14. Yang, H.; Pan, L.; Han, Y.; Ma, L.; Li, Y.; Xu, H.; Zhao, J. A visual water vapor photonic crystal sensor with PVA/SiO2 opal structure. *Appl. Surf. Sci.* **2017**, *423*, 421–425. [CrossRef]
15. Li, Y.; Deng, C.; Yang, M. A novel surface acoustic wave-impedance humidity sensor based on the composite of polyaniline and poly(vinyl alcohol) with a capability of detecting low humidity. *Sens. Actuators B Chem.* **2012**, *165*, 7–12. [CrossRef]
16. Bhadra, J.; Popelka, A.; Abdulkareem, A.; Lehocky, M.; Humpolicek, P.; Al-Thani, N. Effect of humidity on the electrical properties of the silver-polyaniline/polyvinyl alcohol nanocomposites. *Sens. Actuators A Phys.* **2019**, *288*, 47–54. [CrossRef]
17. Mahapure, P.D.; Gangal, S.A.; Aiyer, R.C.; Gosavi, S.W. Combination of polymeric substrates and metal–polymer nanocomposites for optical humidity sensors. *J. Appl. Polym. Sci.* **2019**, *136*, 47035. [CrossRef]
18. Christova, D.; Ivanova, S.; Ivanova, G. Water-soluble temperature-responsive poly(vinyl alcohol-*co*-vinyl acetal)s. *Polym. Bull.* **2003**, *50*, 367–372. [CrossRef]
19. Farmer, P.H.; Jemmott, B.A. Polyvinyl Acetal Adhesives. In *Handbook of Adhesives*; Skeist, I., Ed.; Springer: Boston, MA, USA, 1990; pp. 423–436.

20. Lazarova, K.; Bozhilova, S.; Christova, D.; Babeva, T. Poly(vinyl alcohol)—Based thin films for optical sensing of humidity. *J. Phys. Conf. Ser.* **2020**, in press.
21. Lazarova, K.; Bozhilova, S.; Ivanova, S.; Christova, D.; Babeva, T. The Influence of Annealing on Optical and Humidity Sensing Properties of Poly(Vinyl Alcohol-Co-Vinyl Acetal) Thin Films. *Proceedings* **2020**, *42*, 16. [CrossRef]
22. Schubert, U. Chemistry and Fundamentals of the Sol–Gel Process. In *The Sol–Gel Handbook: Synthesis, Characterization, and Applications*, 1st ed.; Levy, D., Zayat, M., Eds.; Wiley-VCH Verlag GmbH & Co.: Weinheim, Germany, 2015; pp. 3–27.
23. Pingan, H.; Mengjun, J.; Yanyan, Z.; Ling, H. A silica/PVA adhesive hybrid material with high transparency, thermostability and mechanical strength. *RSC Adv.* **2017**, *7*, 2450–2459. [CrossRef]
24. Lazarova, K.; Vasileva, M.; Marinov, G.; Babeva, T. Optical characterization of sol-gel derived Nb_2O_5 thin films. *Opt. Laser Technol.* **2014**, *58*, 114–118. [CrossRef]
25. Lazarova, K.; Christova, D.; Georgiev, R.; Georgieva, B.; Babeva, T. Optical Sensing of Humidity Using Polymer Top-Covered Bragg Stacks and Polymer/Metal Thin Film Structures. *Nanomaterials* **2019**, *9*, 875. [CrossRef] [PubMed]
26. Lazarova, K.; Georgiev, R.; Christova, D.; Babeva, T. Polymer Top-Covered Bragg Reflectors as Optical Humidity Sensors. *Proceedings* **2019**, *3*, 12. [CrossRef]

© 2020 by the authors. Licensee MDPI, Basel, Switzerland. This article is an open access article distributed under the terms and conditions of the Creative Commons Attribution (CC BY) license (http://creativecommons.org/licenses/by/4.0/).

Article

Thin Films of Tolane Aggregates for Faraday Rotation: Materials and Measurement

Maarten Eerdekens [1], Ismael López-Duarte [2], Gunther Hennrich [2,*] and Thierry Verbiest [1,*]

1. Department of Chemistry, University of Leuven, Celestijnenlaan 200D, 3001 Leuven, Belgium; maarten.eerdekens@kuleuven.be
2. Department of Organic Chemistry, Universidad Autónoma de Madrid, Cantoblanco, 28049 Madrid, Spain; ismael.lopez@uam.es
* Correspondence: gunther.hennrich@uam.es (G.H.); thierry.verbiest@kuleuven.be (T.V.)

Received: 11 September 2019; Accepted: 12 October 2019; Published: 16 October 2019

Abstract: We present organic, diamagnetic materials based on structurally simple (hetero-)tolane derivatives. They form crystalline thin-film aggregates that are suitable for Faraday rotation (FR) spectroscopy. The resulting new materials are characterized appropriately by common spectroscopic (NMR, UV-Vis), microscopy (POM), and XRD techniques. The spectroscopic studies give extremely high FR activities, thus making these materials promising candidates for future practical applications. Other than a proper explanation, we insist on the complexity of designing efficient FR materials starting from single molecules.

Keywords: faraday rotation; thin films; magneto-optics; organic material; tolane derivatives

1. Introduction

Faraday rotation (FR) is a magneto-optic (MO) effect that was discovered more than a century ago [1]. It is the rotation of the plane of polarization in the presence of a longitudinal magnetic field, and the rotation angle θ can be described by $\theta = VBL$ with the angle of polarization rotation, V the Verdet constant, B the magnetic field parallel to the propagation of light, and L the path length. Applications of FR are of practical relevance for magnetic field sensors, wave guiding, fiber-optics, etc. [2–4]. Traditionally the field of magneto-optics has been dominated by inorganic materials or radical species [5–8]. Only recently have diamagnetic organic materials emerged as novel FR supplies [9–14].

Although the exact origin of Faraday rotation in organic molecules is currently unknown, different research groups have dedicated their efforts to designing new organic materials for FR applications. Current experiments reported in the literature clearly suggest that molecular conjugation and π-stacking are crucial factors to obtain very strong FR. Furthermore, for organic diamagnetic materials, it became evident that the macroscopic order of the bulk material is crucial for its optical and MO activity [9–14]. It is this duality of molecular vs. macroscopic material, i.e., intra- vs. intermolecular processes, that complicates a rational correlation of the observed magnetic effects with the nature of molecular units and supramolecular aggregates. We have recently shown how the structural simplification of the molecular units (from trigonal to linear) has led to an increase in the FR activity of the respective thin-film materials [15]. Nonetheless, one has to keep in mind the macroscopic structure of the respective aggregates. A decisive requirement is the capacity to form quality thin films from molecular units or aggregates, either crystalline [16] or liquid crystalline [17]. It was shown that long-range electron movement along columnar supramolecular aggregates leads to a dramatically increased Faraday response [18].

2. Materials and Methods

In the present work, we present tolane structures that form crystalline thin films and assess their MO activity by FR spectroscopy (Figure 1).

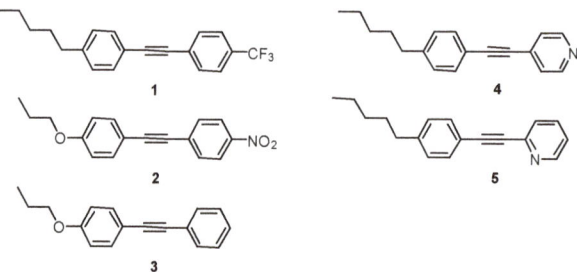

Figure 1. Tolanes (**1**–**3**) and *N*-Hetero-tolane derivatives **4** and **5**.

The diphenyltolanes **1**–**3** present a conventional donor-π-acceptor system. Compounds **2** and **3** have been studied previously for their second-order NLO properties [19].

2.1. Bulk Properties

Tolane **1** was synthesized following the literature procedure and was obtained in a 72% yield [20]. Characterization: ^{1}H NMR (400 MHz, CDCl$_3$) δ_H 7.56 (m, 4H), 7.47 (d, *J* = 8.1 Hz, 2H), 7.17 (d, *J* = 8.1 Hz, 2H), 2.60 (t, *J* = 7.5 Hz, 2H), 1.61 (q, *J* = 7.5 Hz, 2H), 1.31 (m, 4H), 0.91 (t, *J* = 6.8 Hz, 3H); ^{13}C NMR (100 MHz, CDCl$_3$) δ_C 144.5, 131.9, 131.9, 130.5, 129.6, 128.7, 127.6, 125.4, 119.9, 92.3, 87.6, 36.1, 31.6, 31.0, 22.7, 14.1. EI$^+$-MS m/z 316 (M+); Anal. calcd. for: C$_{20}$H$_{19}$F$_3$: %C, 75.93; %H, 6.05; found: %C, 75.88; %H, 5.97.

Differential scanning calorimetry (DSC) analysis revealed the existence of two main crystalline polymorphs melting at 69.7 and 71.3 °C, respectively. These transitions stabilized after two heating–cooling cycles (Figure 2).

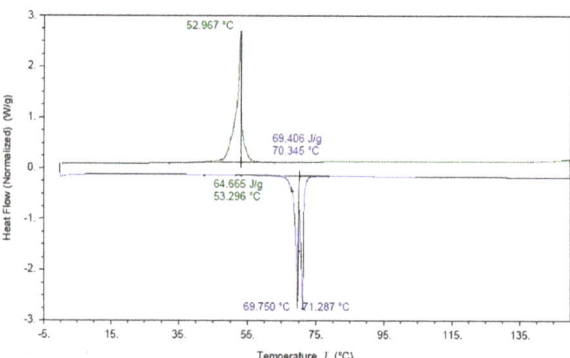

Figure 2. Differential scanning calorimetry (DSC) of **1**, second heating (green) and cooling (blue) cycle.

The *N*-pyridyl-tolanes **4** and **5** were solid (**4**) or liquid (**5**) at room temperature, respectively. However, upon protonation (with hydrochloric or terephthalic acid), both formed crystalline solids. In addition, solid halogen bond complexes were obtained from **4** and **5** with suitable halogen bond donors. The terephthalate complex of protonated **4** was liquid crystalline over a wide temperature range. Hence, the material can be processed and measured conveniently in a conventional liquid crystal (LC) cell.

2.2. Spectroscopy and Thin Film Preparation

UV-Vis Spectra were measured in chloroform at a concentration of 1.5×10^{-4} mol/L using a Perkin-Elmer Lambda 900 spectrophotometer (Norwalk, CT, USA). To measure the Faraday rotation as well as polarized optical microscopy (POM) [21], the materials were placed in LC cells with a 3 μm gap. To fill the cells with the organic materials, a heating plate heated the cells to a temperature of 5 to 10 °C higher than the melting temperature of the desired molecules. A small amount of material was deposited next to the gap. It subsequently melted, and entered the cell through capillary action. After the LC cell had been filled and cooled, a homemade heating and cooling device reheated the filled LC cells to 5 °C above the melting temperature of the organic material, and cooled the samples to r.t. by 0.1 °C/min. An Olympus microscope was used for obtaining POM images. Faraday rotation spectra were collected using a photoelastic modulation magneto-optical setup described by Vandendriessche et al. from 350 to 700 nm, every 2 nm [4]. The optical rotation was measured at varying magnetic field from 0 to 0.5 T. A blank was also measured to nullify the effects of the glass. Using linear regression, the magnetic rotation was calculated from the slope. The Verdet constant (°/Tm) was then calculated by dividing the magnetic rotation by the thickness of the sample inside the LC cell, i.e., 3 μm. Smoothing of the curves was done using Savitsky–Golay method in Origin. We confirmed that the sample was in the plane was isotropic by measuring at different azimuthal angles of the samples at 400 nm (Figure S1).

3. Results and Discussion

Compounds **1–4** were crystalline solids at room temperature. The powder X-ray diffractograms were measured on a Malvern PANalytical Empyrean system, with a Cu K-α source with a wavelength of 1.5406 Å, measuring with a PIXcel3D detector (Malvern analytical, Eindhoven, Noord-Brabant, The Netherlands). The measurements were done at room temperature. The XRD (and POM) measurements confirmed the crystal nature of the samples (Figure S2).

The absorbance spectra of **1** and **2** showed a maximum absorbance of around 350 nm with a corresponding high FR of several hundred thousand °/Tm. This is not surprising since the FR response is enhanced near resonances. However, what is surprising is that even far away from resonance strong Faraday rotation was observed. For example, for compound 1, the Verdet constant in the wavelength region 525 to 700 nm nanometers was still on the order of 50,000 to 70,000°/Tm, while compound 2 exhibits Verdet constants over 150,000°/Tm around 500 nm. Similar behavior has been observed for other crystalline acetylenes [5]. Both molecules had an electron acceptor group (–NO$_2$ and –CF$_3$) within a conjugated π-system. They showed very similar FR spectra with several peaks and valleys in the visible part of the spectrum (Figure 3A). Tolane 2 exhibited a higher Faraday response over most part of the spectrum (Figure 3B).

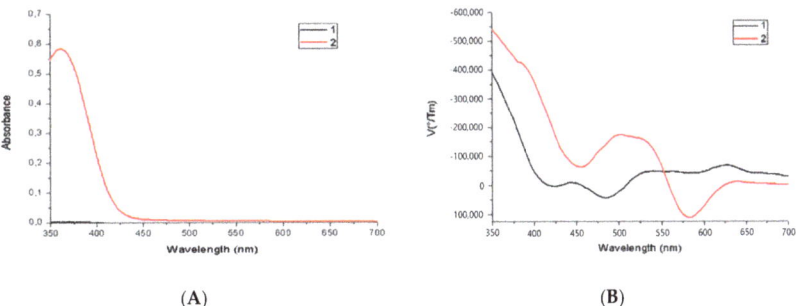

Figure 3. (**A**) and Faraday rotation spectrum (**B**) of **1** and **2**.

Our results for compounds **4** and **5**, nicely illustrate the importance of their macroscopic structure. Compound **5** was an isotropic liquid at r.t. temperature, while compound 4 was solid at room temperature. Therefore and in agreement with earlier findings, **4** exhibited a Faraday response that was orders of magnitude higher over the entire °/Tm, even outside regions of absorption. For 5, we measured a featureless and low-intensity Faraday spectrum that gradually decreased towards the IR region (Figure 4B). Note also that the Faraday spectrum of 4 resembles that of **1** and **2** with a maximum Verdet constant near resonance and a very feature-rich shape in the visible part of the spectrum. This seems to indicate that molecular structure does have an impact on the shape of the Faraday spectrum. Moreover, the supramolecular organization (either in a crystalline or liquid form) is a necessary requirement to observe strong FR activity.

Of all the samples, the strong increase in FR towards the UV part of the spectrum was due to the presence of the absorption band as well as the wavelength dependency of the Verdet constant ($V \sim \lambda^{-2}$). We do not know the origin of the peaks and valleys, but recent work on Faraday rotation in other organic molecules suggests crediting them to the presence of spin-forbidden or hidden singlet and/or triplet states [4]. The non-substituted diphenylacetylene 3 exhibited too much birefringence and scattering to perform reliable measurements. Its UV-Vis absorption spectrum can be found in the supporting information (Figure S3). POM images can be found in the supporting information (Figure S4).

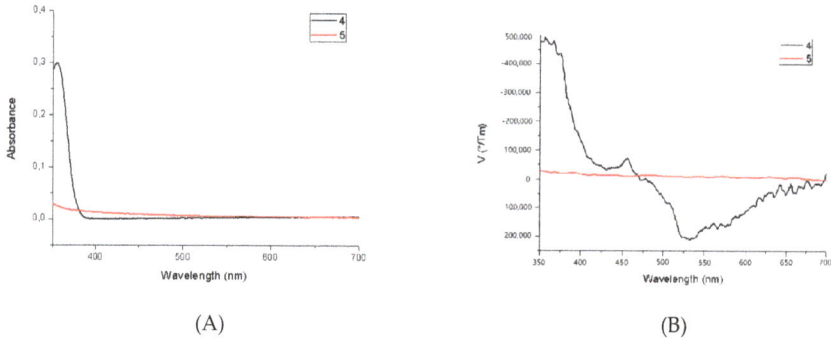

Figure 4. UV-Vis (**A**) and Faraday rotation spectrum (**B**) of 4 and 5.

4. Conclusions

We have investigated the FR response of different phenylacethylene derivatives. It is clear that macroscopic organization within the bulk material is a key factor in obtaining a high Faraday response. The molecular structure dictates the macroscopic organization of the building units, which, in consequence, determines the FR response of the resulting bulk material. The detailed shape of the Faraday spectrum is a result of this (i.e., the molecular structure), but is in itself not sufficient to create a high Faraday response. Once again, we have to emphasize the aforementioned duality (single molecule vs. macroscopic bulk) that interferes with the design of efficient FR materials.

The tolanes that were crystalline at room temperature showed very high Verdet constants—much higher than typically observed for diamagnetic materials—in regions of the spectrum where there is no optical absorption, making them potentially useful for applications.

Supplementary Materials: The following are available online at http://www.mdpi.com/2079-6412/9/10/669/s1, Figure S1: Verdet constant measurements at 400 nm: Verdet constant of samples 1, 2, 4 and 5 turned azimuthal 0°, 30°, 60° and 90°. The Verdet constant was measured at 400 nm. No dependence of Verdet constant on the rotation of azimuthal angle was observed; Figure S2: X-ray diffractograms: tolanes (1–3) and n-hetero-tolane derivatives 4; Figure S3: UV-Vis absorbance spectrum of the unsubstituted diphenylacetylene (3); Figure S4: Polarized optical microscopy: polarized optical microscopy images of the materials in the LC cells.

Author Contributions: Conceptualization, T.V. and G.H.; methodology, M.E.; validation, M.E.; formal analysis, M.E.; investigation, M.E.; Resources: LC: T.V. and M.E., Synthesis, I.L.-D. and G.H.; data curation, M.E.; writing—original draft preparation, M.E. and T.V.; writing—review and editing, M.E., T.V. and G.H.; visualization, M.E. and G.H.; supervision, M.E. and G.H. and T.V., project administration, M.E. and G.H.; funding acquisition, G.H and T.V. All authors have given approval to the final version of the manuscript.

Funding: This research was funded by the Spanish government (MINECO, project CTQQ2016-7557-R) and the KULeuven (C1 project).

Acknowledgments: The authors would like to thank B. Goderis for using his POM equipment and M. Rouffaers and L. Clinckemalie for the X-ray data analysis.

Conflicts of Interest: The authors declare no conflict of interest.

References

1. Faraday, M. *Experimental Researches in Electricity*; R. Taylor & W. Francis: London, UK, 1839; Volume III.
2. Jalas, D.; Petrov, A.; Eich, M.; Freude, W.; Fan, S.; Yu, Z.; Baets, R.; Popović, M.; Melloni, A.; Joannopoulos, J.D. What is—and what is not—an optical isolator. *Nat. Photonics* **2013**, *7*, 579. [CrossRef]
3. Schmidt, M.A.; Wondraczek, L.; Lee, H.W.; Granzow, N.; Da, N.; Russell, P.S.J. Complex faraday rotation in microstructured magneto-optical fiber waveguides. *Adv. Mater.* **2011**, *23*, 2681–2688. [CrossRef] [PubMed]
4. Amirsolaimani, B.; Gangopadhyay, P.; Persoons, A.P.; Showghi, S.A.; LaComb, L.J.; Norwood, R.A.; Peyghambarian, N. High sensitivity magnetometer using nano-composite polymers with large magneto-optic response. *Opt. Lett.* **2018**, *43*, 4615–4618. [CrossRef] [PubMed]
5. Hsiao, Y.-C.; Wu, T.; Li, M.; Hu, B. Magneto-optical studies on spin-dependent charge recombination and dissociation in perovskite solar cells. *Adv. Mater.* **2015**, *27*, 2899–2906. [CrossRef] [PubMed]
6. Lim, C.-K.; Cho, M.J.; Singh, A.; Li, Q.; Kim, W.J.; Jee, H.S.; Fillman, K.L.; Carpenter, S.H.; Neidig, M.L.; Baev, A. Manipulating Magneto-Optic Properties of a Chiral Polymer by Doping with Stable Organic Biradicals. *Nano Lett.* **2016**, *16*, 5451–5455. [CrossRef] [PubMed]
7. Ma, J.; Hu, J.; Li, Z.; Nan, C.-W. Recent progress in multiferroicmagnetoelectric composites. *Adv. Mater.* **2011**, *23*, 1062–1087. [CrossRef] [PubMed]
8. Wang, P.; Lin, S.; Lin, Z.; Peeks, M.D.; Van Voorhis, T.; Swager, T.M. A semiconducting conjugated radical polymer: Ambipolar redox activity and faraday effect. *J. Am. Chem. Soc.* **2018**, *140*, 10881–10895. [CrossRef] [PubMed]
9. Gangopadhyay, P.; Foerier, S.; Koeckelberghs, G.; Vangheluwe, M.; Persoons, A.; Verbiest, T. Efficient Faraday rotation in conjugated polymers. *Proc. SPIE* **2006**, *6331*, 63310Z.
10. Koeckelberghs, G.; Vangheluwe, M.; Doorsselaere, K.V.; Robijns, E.; Persoons, A.; Verbiest, T. Regioregularity in poly(3-alkoxythiophene)s: Effects on the Faraday rotation and polymerization mechanism. *Macromol. Rapid Commun.* **2006**, *27*, 1920–1925. [CrossRef]
11. Vandendriessche, S.; Van Cleuvenbergen, S.; Willot, P.; Hennrich, G.; Srebro, M.; Valev, V.K.; Koeckelberghs, G.; Clays, K.; Autschbach, J.; Verbiest, T. Giant Faraday rotation in mesogenic organic molecules. *Chem. Mater.* **2013**, *25*, 1139–1145. [CrossRef]
12. Wang, P.; Jeon, I.; Lin, Z.; Peeks, M.D.; Savagatrup, S.; Kooi, S.E.; Van Voorhis, T.; Swager, T.M. Insights into Magneto-Optics of Helical Conjugated Polymers. *J. Am. Chem. Soc.* **2018**, *140*, 6501–6508. [CrossRef] [PubMed]
13. Gangopadhyay, P.; Koeckelberghs, G.; Persoons, A. Magneto-optic properties of Regioregular Polyalkylthiophenes. *Chem. Mater.* **2011**, *23*, 516–521. [CrossRef]
14. Thompson, J.R.; Ovens, J.S.; Williams, V.E.; Leznoff, D.B. Supramolecular assembly of Bis(benzimidazole)pyridine: An extended anisotropic ligand for highly birefringent materials. *Chem. Eur. J.* **2013**, *19*, 16572–16578. [CrossRef] [PubMed]
15. Vleugels, R.; de Vega, L.; Brullot, W.; Verbiest, T.; Gómez-Lor, B.; Gutierrez-Puebla, E.; Hennrich, G. Magneto-optical activity in organic thin film materials. *Smart Mater. Struct.* **2016**, *25*, 12LT01. [CrossRef]
16. Desiraju, G.R. *Perspectives in Supramolecular Chemistry: The Crystal as Supramolecular Entity*; John Wiley & Sons: New York, NY, USA, 2007.
17. Wang, L.; Li, Q. Stimuli-directing self-organized 3D liquid-crystalline nanostructures: From materials design to photonic applications. *Adv. Funct. Mater.* **2016**, *36*, 10–28. [CrossRef]

18. Vleugels, R.; Steverlynck, J.; Brullot, W.; Koeckelberghs, G.; Verbiest, T. Faraday rotation in discotic liquid crystals by long range electron movement. *J. Phys. Chem. C* **2019**, *123*, 9382–9387. [CrossRef]
19. Hennrich, G.; Murillo, M.T.; Prados, P.; Al-Saraierh, H.; El-Dali, A.; Thompson, D.W.; Collins, J.; Georghiou, P.E.; Teshome, A.; Asselberghs, I. Alkynyl Expanded Donor–Acceptor Calixarenes: Geometry and Second-Order Nonlinear Optical Properties. *Chem. Eur. J.* **2007**, *13*, 7753–7761. [CrossRef] [PubMed]
20. Sonogashira, K. Development of Pd–Cu catalyzed cross-coupling of terminal acetylenes with sp^2-carbon halides. *J. Organomet. Chem.* **2002**, *653*, 45–49. [CrossRef]
21. Outram, B. Polarising Optical Microscopy. In *Liquid Crystals*; IOP Publishing: Bristol, UK, 2018.

© 2019 by the authors. Licensee MDPI, Basel, Switzerland. This article is an open access article distributed under the terms and conditions of the Creative Commons Attribution (CC BY) license (http://creativecommons.org/licenses/by/4.0/).

Article

Thermal Stability of CrWN Glass Molding Coatings after Vacuum Annealing

Xinfang Huang [1], Zhiwen Xie [1,*], Kangsen Li [2], Qiang Chen [3], Yongjun Chen [1,4] and Feng Gong [2,*]

1. Liaoning Key Laboratory of Complex Workpiece Surface Special Machining, University of Science and Technology Liaoning, Anshan 114051, China; xfgg527@163.com (X.H.); chenyongjun-net@163.com (Y.C.)
2. Shenzhen Key Laboratory of Advanced Manufacturing Technology for Mold and Die, Shenzhen University, Shenzhen 518060, China; likangsenszu@163.com
3. Southwest Technology and Engineering Research Institute, Chongqing 400039, China; 2009chenqiang@163.com
4. Key Laboratory of Materials Modification by Laser, Ion and Electron Beams (Dalian University of Technology), Ministry of Education, Dalian 116000, China
* Correspondence: xzwustl@126.com (Z.X.); gongfeng186@163.com (F.G.); Tel.: +86-412-592-9746 (Z.X.); +86-755-26558509 (F.G.)

Received: 27 December 2019; Accepted: 21 February 2020; Published: 25 February 2020

Abstract: CrWN glass molding coatings were deposited by plasma enhanced magnetron sputtering (PEMS). The microstructure and thermal stability of these coatings were investigated by X-ray diffraction, X-ray photoelectron spectroscopy, scanning electron microscope, transmission electron microscope, atomic force microscope and nanoindentation tests. The as-deposited coating exhibited an aggravated lattice expansion resulting in a constant hardness enhancement. The vacuum annealing induced surface coarsening and the spinodal decomposition of the coating accompanied by the formation of nm-sized c-CrN, c-W_2N, and h-WN domains. The annealed coating with low W content had mainly a face-centered cubic (f.c.c) matrix, strain fields caused by lattice mismatch caused hardness enhancement. Following an increase in W content, the annealed coating showed a mixed face-centered cubic (f.c.c) and hexagonal close-packed (h.c.p) matrix. The large volume fraction of h-WN phases seriously weakened the coating strengthening effect and caused an obvious drop in hardness.

Keywords: CrWN coatings; microstructure evolution; spinodal decomposition; thermal stability; hardness; plasma enhanced magnetron sputtering

1. Introduction

Optical glass lenses have a broad range of applications (e.g., telescopes, cameras, medical equipment, and high-power lasers) because of their good refractive index, excellent chemical stability, light permeability, and high image quality [1–3]. Conventional manufacturing technologies include several techniques (e.g., precision grinding, and ultra-precision lapping and polishing), but these methods are time-consuming and difficult to use for the fabrication of complex shapes [4]. Precision molding, however, has low cost, high efficiency, can be applied to net forming, and is environmentally friendly. These characteristics have attracted great interest from optical manufacturers [5–7]. Unfortunately, the adhesive wear of glass blanks greatly degrades the molding life of the forming die and eventually decreases the quality of the optical glass component surface [8,9].

It has been reported that protective coatings can effectively improve the anti-sticking and anti-wear performance of forming dies [10–18]. Precious metal coatings are widely used in molding manufacturing due to their excellent anti-sticking and anti-oxidation properties [10–12], but they are also expensive for their wide industrial applications. Diamond-like carbon (DLC) films exhibit good self-lubrication and anti-wear performance, however, their poor thermal stability greatly limits any molding application [13,14]. Recently, the application of transition metal nitride coatings to glass molding has received great interest due to the favorable chemical inertness and anti-wear properties of these materials [15–18]. Most current studies focus on the anti-sticking and anti-oxidation properties of transition metal nitride coatings. WCrN coatings have been reported to have good anti-sticking properties at 400 °C, but also exhibited poor anti-sticking performances at 500 °C due to the formation of oxides [19]. CrWN coatings have been reported to experience severe mechanical degradation and coarsening in high temperature nitrogen atmosphere due to the formation of WO_3 phases [20].

The life of glass molding coatings is determined mainly by their anti-sticking and anti-oxidation properties, as well as by their thermal stability. However, relatively little work has been published on the thermal stability of this type of coatings. In this study, CrWN coatings with different W contents are synthesized using PEMS. Detailed characterizations of these coatings were performed to study the evolution of their microstructures, as well as the mechanical properties of the as-deposited coatings after vacuum annealing. The potential effect of the W content on the thermal stability of the annealed coatings will be discussed systematically in the following sections.

2. Materials and Methods

CrWN coatings having different W contents were synthesized by PEMS [21]. Pure W (99.9%) and Cr (99.6%) were used as sputtering targets. A silicon wafer and cemented carbides (WC-8 wt % Co) were used for the substrate. The samples were mechanically polished using a diamond paste and ultrasonically washed using pure ethanol. High purity nitrogen and argon were utilized for the working atmosphere. The samples were placed on a rotated holder, moving at a rotating speed of 2 rpm. The chamber was evacuated to a base pressure of 5×10^{-3} Pa and then heated to 300 °C. Prior to coating deposition, the samples were cleaned by Ar^+ sputtering to remove any residual pollution and the native oxides. The sputtering parameters were as follows: bias voltage of −120 V, argon flow of 140 sccm, sputtering time 60 min. The CrWN coatings were synthesized in a mixed atmosphere of nitrogen and argon. The chemical composition of the as-deposited coating was varied by adjusting the powers of the Cr and W targets. In this case, the deposition parameters were displayed as follows: bias voltage of −50 V, argon flow 100 sccm, nitrogen flow 100 sccm, and deposition time 100 min. The deposition parameters are listed in Table 1.

Table 1. Experimental details of as-deposited coatings.

Samples	Power of Cr Target (W)	Power of W Target (W)	Time (min)
C1	5000	4000	65
C2	2700	4000	100
C3	1600	5100	100

The compositions of the coatings were determined by an energy dispersive spectrometer (EDS) combined with a scanning electron microscope (SEM, Zeiss ∑IGMA HD, Oberkochen, Germany). Moreover, the crystalline structures of the coatings were investigated by X-ray diffraction (XRD, X' Pert Powder, Malvern Panalytical, Malvern, UK) using a Cu Kα radiation source with parallel beam. The incident angle was of 1°, while the diffraction angle was scanned from 20° to 90°. Additionally, X-ray photoelectron spectroscopy (XPS, ThermoFisher, K-Alpha+, Waltham, MA, USA) was employed to detect the chemical states of the coatings. The data were collected after 30 s from etching to remove any contaminants adsorbed on the coating surface. A SEM (Zeiss ∑IGMA HD) was employed to characterize the surface morphology of the coatings, while their microstructure was investigated with a transmission electron microscope (TEM, FEI Titan Cubed Themis G2 300, ThermoFisher, Waltham, MA, USA). Moreover, atomic force microscopy (AFM, Oxford MFP-3DInfinity, Abingdon, UK) was employed to evaluate the surface roughness of the coatings: the scanning area of each image was of 10×10 μm. The mechanical properties of the coatings were evaluated by using a nm-indentation system (Hystron TI950, Bruker, Billerica, MA, USA): five indentations were performed on each sample; additionally, the hardness and elastic modulus selected at a depth of corresponding to 1/10 of the coating thickness, as to minimize the negative effect of the substrate. Finally, vacuum annealing was conducted in a heat treatment furnace at a pressure of 0.1 Pa (RTP-500, Beijing Ruiyisi Technology Co.LTD, Beijing, China), the annealing temperature and time were of 650 °C and 300 min, respectively.

3. Results

Table 2 summarizes the chemical compositions of the Cr-W-N coatings synthesized using different sputtering powers. The Cr content varied from 13.9 ± 0.7 at % to 35.9 ± 2.3 at %, the W content varied from 21.8 ± 5.0 at % to 42.1 ± 6.9 at %, and the N content varied from 40.2 ± 4.8 at % to 43.1 ± 4.4 at %. The oxygen content fluctuated between 0.9 ± 0.4 and 2.1 ± 0.6 at %.

Table 2. Chemical compositions of as-deposited coatings.

Sample	Cr at %	W at %	N at %	O at %
C1	35.9 ± 2.3	21.8 ± 5.0	40.2 ± 4.8	2.1 ± 0.6
C2	26.4 ± 1.4	31.9 ± 6.1	40.3 ± 4.5	1.4 ± 0.5
C3	13.9 ± 0.7	42.1 ± 6.9	43.1 ± 4.4	0.9 ± 0.4

Figure 1 shows the XRD patterns of the as-deposited coatings. The main diffraction peaks were observed at diffraction angles of 37.21°, 43.27°, 62.85°, 75.64° and 78.73°, corresponding to the (111), (200), (220), (311) and (222) planes of both the c-CrN and c-W_2N phases [22]. The value of these diffraction peaks slightly shifted to lower angles with the increase of W content. This change likely originated from lattice expansion due to the solid solution of W atoms [20,23,24]. In addition, the coatings exhibited a (111) preferential orientation, although the intensity of the strong (111) peak was considerably lower as the W content is increased from 21.8 ± 5.0 at.% to 42.1 ± 6.9 at %, The results are in accordance with those of previous studies, indicating that the doping of a small amount of W to CrN can induce an evolution of the texture from a preferential orientation of (111) to one of (200) [20]. Moreover, the weak (222) peak is very wide and close to the diffraction angle of (311) peak, leading to the asymmetry of the (311) peak.

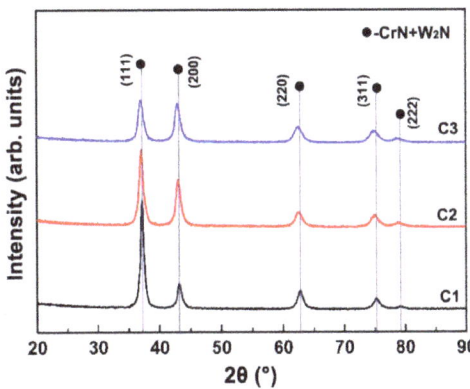

Figure 1. XRD patterns of as-deposited coatings.

Figure 2 shows the XPS spectra of Cr2p, W4f, N1s and O1s in sample C2. As shown in Figure 2a, the binding energies of the $Cr2p_{3/2}$ and $Cr2p_{1/2}$ peaks centered at 574.7 eV and 583.4 eV, which correspond to the Cr-N state [25,26]. Meanwhile, the binding energies of $Cr2p_{3/2}$ and $Cr2p_{1/2}$ peaks at 576.6 and 586.3 eV correspond to the Cr–O states [27]. The binding energies of $W4f_{7/2}$ and $W4f_{5/2}$ peaks located at 31.8 eV and 33.9 eV (Figure 2b), which are associated with W-N binding state [28,29]. The binding energies of W4f$_{7/2}$ and W4f$_{5/2}$ peaks at 35.6 eV and 37.8 eV are assigned to the W–O state [30]. Figure 2c displays that the binding energies of N1s peaked at 396.9 eV, 397.7 eV and 399.8 eV, which can be assign to Cr-N and W-N binding states, respectively [25,28]. The O1s spectrum in Figure 2d shows that the binding energies of the O1s peaked at 530.1, 530.7 and 531.8 eV, which can be assign to Cr–O, W–O and H–O states, respectively [31–33].

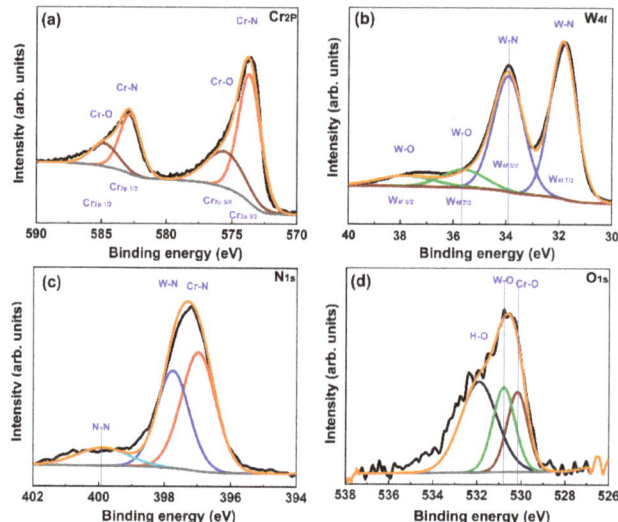

Figure 2. XPS spectra of sample C2: (**a**) Cr2p, (**b**) W4f, (**c**) N1s, (**d**) O1s.

Figure 3 shows the surface images of the as-deposited coatings. Sample C1 exhibited a granular morphology, as shown in Figure 3a, plenty of micropores were distributed around grain boundaries, indicating a loose and coarse growth structure. The surface features changed considerably with an increase of the W content. As shown in Figure 3b, sample C2 exhibited a dense surface morphology, plenty of nm-sized grains were closely packed in large cluster particles. Sample C3 also showed a dense surface (Figure 3c). There were plenty of cluster particles consisting of numerous fine grains, but the sizes of such cluster particles decreased considerably in comparison to sample C2.

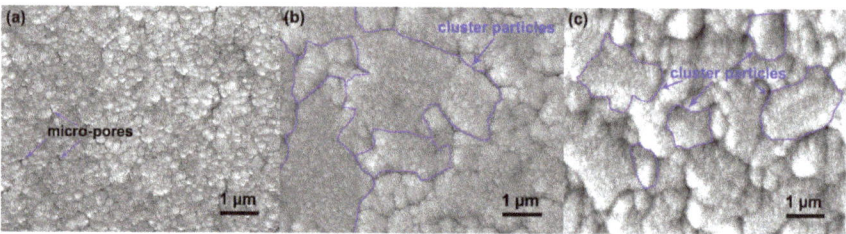

Figure 3. Surface SEM images of as-deposited coatings: (**a**) C1, (**b**) C2, (**c**) C3.

Figure 4 shows the cross-section TEM images of sample C2. The dark field TEM image in Figure 4a displayed a distinct columnar grain morphology. Moreover, a polycrystalline cubic structure with (111), (200), (220), and (311) reflections was identified according to the selected area diffraction pattern (SADP). The HAADF STEM image in Figure 4b showed an obvious two-layered structure with alternate bright (W$_2$N) and dark (CrN) contrast according to the element mapping image. These multilayer structure mainly originated from the rotation of sample holder during the coating deposition. Meanwhile, these nano-multilayers exhibited a typical coherent epitaxial growth mode (Figure 4c). The excellent coherence relations were clearly kept between W_2N and CrN phases, as shown in Figure 4d, the d-spacing values were 0.145 nm for the CrN phase and 0.146 nm for the W_2N phase. These similar d-spacing values were assign to the (220) planes. Additionally, plenty of W atoms were dissolved in the CrN matrix, and some Cr atoms were also identified in the W_2N matrix. Both of which indicated that a significant solid solution effect occurred during the coating deposition.

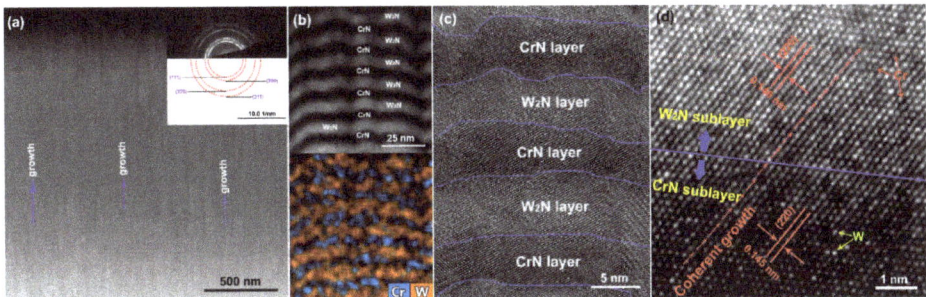

Figure 4. Cross-section TEM images of sample C2: (**a**) TEM image with corresponding SADP, (**b**) HAADF STEM image and element mapping, (**c**,**d**) HRTEM images.

Figure 5 shows the AFM images of the as-deposited coatings. Here sample C1 showed a relatively rough surface with a high Ra value of 10.841 nm (Figure 5a). The roughness value decreased considerably with the increase in W content. Figure 5b shows a smooth surface for sample C2, having a low Ra value of 2.417 nm. Figure 5c indicates a similar Ra value for sample C3 (2.698 nm). Based on the XRD, SEM, TEM, and AFM results, we can infer that the CrWN coating has an improved surface quality, although it undergoes aggravated lattice expansion under increasing of W content.

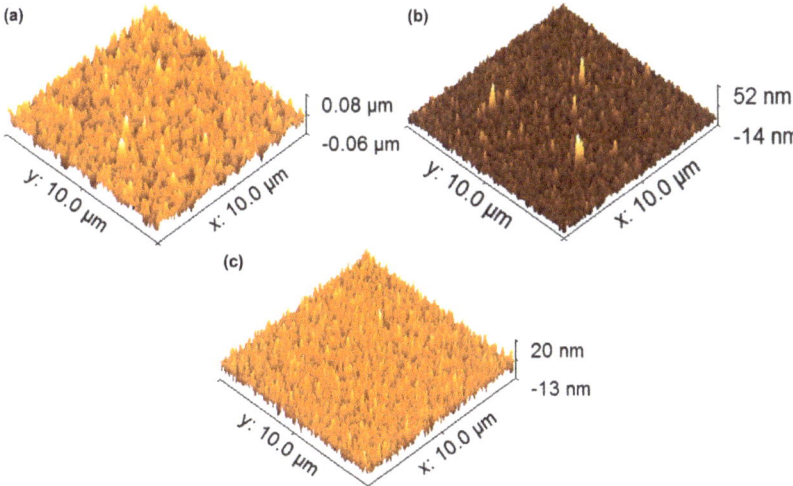

Figure 5. AFM images of as-deposited coatings: (**a**) C1, (**b**) C2, (**c**) C3.

Figure 6 shows the XRD patterns of the vacuum annealed coatings. The annealed samples C1 and C2 exhibited similar phase structures to those of the as-deposited coatings. Strong diffraction peaks of mixed c-CrN and c-W_2N phases can be seen in the XRD patterns. Simultaneously, weak peaks of h-WN phases can be seen in these annealed coatings, implying a slight phase decomposition. However, the phase decomposition is aggravated significantly with the increase of W content. Apart from the original diffraction peaks of the mixed c-CrN and c-W_2N phases, the diffraction intensity of the h-WN peak was found to increase significantly in the annealed sample C3, indicating the formation of a large volume fraction of h-WN phases.

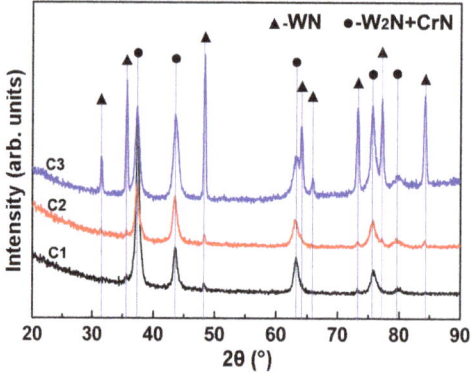

Figure 6. XRD patterns of vacuum annealed coatings.

Figure 7 shows the XRD patterns and cross-section TEM images of as-deposited and annealed sample C2. Compared with the as-deposited coating, the annealed coating remained stable structure of CrN and W_2N phases, but their diffraction peaks obviously shifted to higher angles, indicating a mitigating lattice expansion. Meanwhile, the as-deposited coating showed clearly two-layered structure with alternate bright (W_2N) and dark (CrN) contrast (Figure 7a). By contrast, the annealed coating exhibited obviously three-layered structure with bright (W_2N), gray (mixed CrN-W_2N), and dark (CrN) contrast, as shown in Figure 7b, a distinct gray diffusion layer forms between CrN and W_2N sublayers. The partial enlarged image in Figure 7b reveals that the excellent coherence relations were kept in the diffusion layer. Apparently, the vacuum annealing effectively drove the decomposition of the supersaturated solid solution resulting in the formation of a diffused solid solution matrix.

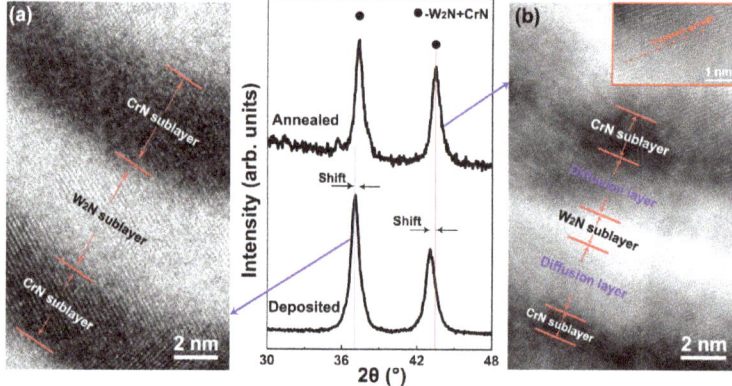

Figure 7. XRD patterns and cross-section TEM images of as-deposited (**a**) and annealed (**b**) sample C2.

Figure 8 shows the surface images and EDS spectra of the vacuum annealed coatings. As shown in Figure 8a, the annealed sample C1 exhibited a granular surface morphology with plenty of newly grown nm particles. The O content in this coating was of 9.6 ± 1.7 at %. Meanwhile, the surface features of the annealed samples C2 and C3 were slightly different from those of the as-deposited coating (Figure 8b,c). These nm-sized grains in fact grew, leading to a visible coarsening of the cluster particles. The O contents were of 7.4 ± 1.3 at % and 12.1 ± 1.7 at % for annealed samples C2 and C3, respectively. The SEM characterizations clearly confirmed that these coatings suffered slight oxidation damages during the vacuum annealing.

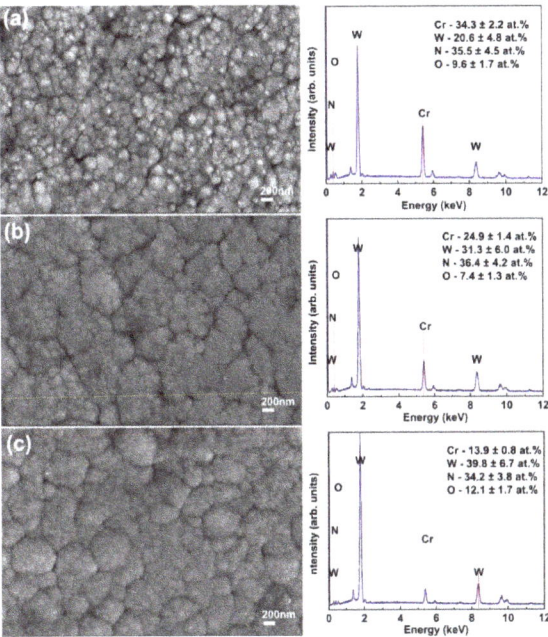

Figure 8. Surface images and EDS spectra of vacuum annealed coatings: (**a**) C1, (**b**) C2, (**c**) C3.

Figure 9 shows the roughness values and AFM images of the as-deposited and annealed coatings. The Ra values of samples C1 and C2 increased from 10.841 to 11.332 nm and from 2.417 to 3.204 nm, respectively; by contrast, the Ra value of sample C3 increased abruptly from 2.698 to 4.945 nm, indicating a severe surface coarsening.

Based on the SEM and AFM characterizations, we found that the annealed coatings underwent various degrees of surface coarsening, which was likely triggered by surface oxidation. The EDS results further showed the occurrence of slight oxidation erosion during vacuum annealing, which consequently resulted in the formation of Cr–W oxides [20]. Although these oxides are too small to be detected by XRD in Figure 6, but they led to severe surface coarsening because of their loose structure [20,34]. The significant differences in Ra value between the annealed coatings having different W content can be likely attributed to the varying composition of the oxide layers. According to the EDS results (Figure 8), the volume fraction of Cr oxides decrease considerably and was replaced by a rising volume fraction of W oxides under increasing W content. Previously, it has been reported that dense Cr oxides can act as protective layers and inhibit O diffusion into the coating [35], whereas the W oxides usually exhibit a more porous structure [34]. Therefore, an increase in W oxides can eventually lead to a constant increase of the Ra values.

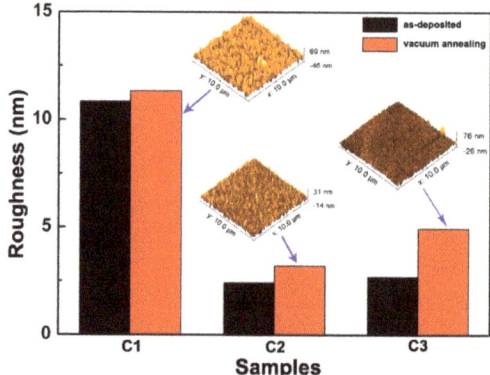

Figure 9. Roughness values and AFM images of as-deposited and annealed coatings: C1, C2, C3.

Figure 10 shows the hardness of the as-deposited and annealed coatings. The as-deposited coatings exhibited a constant hardness enhancement under increasing W content. The hardness was 12.9 GPa for sample C1, and slightly increased to 13.7 GPa for sample C2. By contrast, sample C3 showed a higher hardness of 15.5 GPa. The vacuum annealing induced a remarkable age-hardening in sample C1 and C2. The annealed sample C1 showed a slight increase in hardness (from 12.9 to 14.8 GPa), whereas that of the annealed sample C2 increased rapidly (from 13.7 to 21.6 GPa). Meanwhile, the annealed sample C3 underwent a serious hardness degradation (from 15.5 to 10.1 GPa).

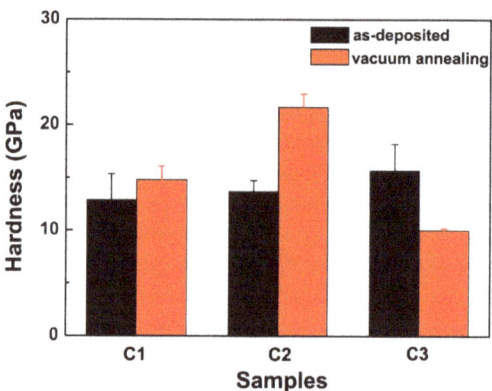

Figure 10. Hardness of as-deposited and annealed coatings.

Based on the XRD and TEM characterizations, it can be inferred that the sputtered atoms or ions triggered an obvious injection effect in the as-deposited coatings, as shown in Figure 4d, a significant solid solution effect appeared in both W_2N and CrN sublayers. According to the previous results [20,23,24], these solid solution atoms induced serious lattice expansion resulting in a visible peak shift as determined from the XRD test (Figure 1). Meanwhile, this lattice expansion was significantly aggravated following an increase in the W content, which provided an obvious strengthening effect and eventually led to a constant hardness enhancement. Moreover, the addition of W to CrN led to an obvious reduction of the particle (i.e., grains or clusters) sizes (Figure 3) and refined their structure, causing a hardness improvement. The annealed coatings with different W contents exhibited significant differences in phase composition, which consequently had a strong effect on their mechanical properties. As determined from the nanoindentation test, the annealing sample C1 and C2 showed a prominent age-hardening, which mainly originated from their microstructure evolution.

As identified in Figures 4d and 7a, these doping atoms induced the formation of supersaturated solid solution in the as-deposited coating, but the supersaturated matrix exhibited a non-uniform distribution resulting in limit strengthening effect. By contrast, vacuum annealing induced spinodal decomposition of supersaturated solid solution to form nm-sized c-CrN, c-W_2N, and h-WN domains (see Figure 6). Although a small amount of h-WN phases was formed, but the annealed coating showed similar face-centred cubic (f.c.c) structure to that of as-deposited coating. According to the previous age-hardening theories [36–39], strain fields, originating from the lattice mismatch, acted as obstacles for the dislocation movement and caused hardness enhancement. Additionally, thermal activation greatly driven the homogenization of the supersaturated matrix, as evidenced in Figure 7b, a distinct diffusion layer formed between CrN and W_2N sublayers. In comparison to the limited strengthening effect in as-deposited coating, the coherent solid solution diffusion matrix provided strong strengthening effect due to the larger mismatch, and consequently resulted in a profound age-hardening in sample C1 and C2. Nevertheless, the annealed sample C3 underwent serious spinodal decomposition accompanied by the formation of a mixed f.c.c and hexagonal close-packed (h.c.p) matrix structure. The large volume fraction of h-WN phases weakened considerably strengthening effect of the coherent interface and eventually led to a drop of the hardness value.

4. Conclusions

This study investigated the microstructure and thermal stability of CrWN glass molding coatings having different W contents after vacuum annealing. The main conclusions are as follows:

- The as-deposited coatings showed columnar structures consisting of multilayer c-CrN and c-W_2N phases. The preferential orientation of these structures varied from (111) to (200), and the surfaces became smoother under increasing W content.
- The as-deposited coatings suffered from aggravated lattice expansion. The grain size and cluster particle sizes showed an obvious decrease under increasing W content. These changes led to a constant hardness enhancement.
- The annealed coatings underwent minor oxidation and varying degrees of surface coarsening. The roughness value gradually increased due to an increasing volume fraction of W oxides.
- The annealed coatings experienced spinodal decomposition while forming nm-sized c-CrN, c-W_2N, and h-WN domains. The volume fraction of the h-WN phase increased significantly with the increase of W content.

The annealed coating under low W content exhibited a significant age-hardening due to the strain fields originating from the lattice mismatch. The annealed coating under high W content suffered from a severe hardness degradation because the large volume fraction of h-WN seriously weakened the interface strengthening effect.

Author Contributions: Conceptualization, Z.X. and F.G.; Methodology, Y.C.; Investigation, X.H. and K.L.; Supervision, Q.C. All authors have read and agreed to the published version of the manuscript.

Funding: This work was supported by National Natural Science Foundation of China (51771087), Liaoning Innovative Talents Support Plan (LR2017052), University of Science and Technology Liaoning Talent Project Grants (601011507-07), and the Innovation Team of Liaoning University of Science and Technology (2017TD04).

Acknowledgments: We sincerely thank the Institute of Nano Surface Engineering of Shenzhen University for their TEM technical supports.

Conflicts of Interest: The authors declare no conflict of interest.

References

1. Zhang, S.; Zhou, L.; Xue, C.; Wang, L. Design and simulation of a superposition compound eye system based on hybrid diffractive-refractive lenses. *J. Appl. Opt.* **2017**, *56*, 7442–7449. [CrossRef] [PubMed]
2. Jiang, C.; Lim, B.; Zhang, S. Three-dimensional shape measurement using a structured light system with dual projectors. *Appl. Opt.* **2018**, *57*, 3983–3990. [CrossRef] [PubMed]

3. Yin, S.; Jia, H.; Zhang, G.; Chen, F.; Zhu, K. Review of small aspheric glass lens molding technologies. *Front. Mech. Eng.* **2017**, *12*, 66–76. [CrossRef]
4. Tang, L.; Zhou, T.; Zhou, J.; Liang, Z.; Wang, X. Research on single point diamond turning of chalcogenide glass aspheric lens. *Procedia CIRP* **2018**, *71*, 293–298. [CrossRef]
5. Tao, B.; He, P.; Shen, L.; Yi, A. Quantitatively measurement and analysis of residual stresses in molded aspherical glass lenses. *Int. J. Adv. Manuf. Tech.* **2014**, *74*, 1167–1174. [CrossRef]
6. Firestone, G.C.; Yi, A.Y. Precision compression molding of glass microlenses and microlens arrays—An experimental study. *Appl. Opt.* **2005**, *44*, 6115–6122. [CrossRef]
7. Yi, A.Y.; Chen, Y.; Klocke, F.; Pongs, G.; Demmer, A.; Grewell, D.; Benatar, A. A high volume precision compression molding process of glass diffractive optics by use of a micromachined fused silica wafer mold and low Tg optical glass. *J. Micromech. Microeng.* **2006**, *16*, 2000–2005. [CrossRef]
8. Rieser, D.; Spieß, G.; Manns, P. Investigations on glass-to-mold sticking in the hot forming process. *J. Non-Crystal. Solids* **2008**, *354*, 1393–1397. [CrossRef]
9. Fischbach, K.D.; Georgiadis, K.; Wang, F.; Dambon, O.; Klocke, F.; Chen, Y.; Allen, Y.Y. Investigation of the effects of process parameters on the glass-to-mold sticking force during precision glass molding. *Surf. Coat. Technol.* **2010**, *205*, 312–319. [CrossRef]
10. Klocke, F.; Bouzakis, K.-D.; Georgiadis, K.; Gerardis, S.; Skordaris, G.; Pappa, M. Adhesive interlayers' effect on the entire structure strength of glass molding tools' Pt-Ir coatings by nano-tests determined. *Surf. Coat. Technol.* **2011**, *206*, 1867–1872. [CrossRef]
11. Zhu, X.-Y.; Wei, J.-J.; Chen, L.-X.; Liu, J.-L.; Hei, L.-F.; Li, C.-M.; Zhang, Y. Anti-sticking Re-Ir coating for glass molding process. *Thin Solid Films* **2015**, *584*, 305–309. [CrossRef]
12. Tseng, S.F.; Lee, C.T.; Huang, K.C.; Chiang, D.; Huang, C.Y.; Chou, C.P. Mechanical properties of Pt-Ir and Ni-Ir binary alloys for glass-molding dies coating. *J. Nanosci. Nanotechnol.* **2011**, *11*, 8682–8688. [CrossRef] [PubMed]
13. Bernhardt, F.; Georgiadis, K.; Dolle, L.; Dambon, O.; Klocke, F. Development of a ta-C diamond-like carbon (DLC) coating by magnetron sputtering for use in precision glass molding. *Materialwissenschaft und Werkstofftechnik* **2013**, *44*, 661–666. [CrossRef]
14. Xie, Z.W.; Wang, L.P.; Wang, X.F.; Huang, L.; Lu, Y.; Yan, J.C. Influence of high temperature annealing on the structure, hardness and tribological properties of diamond-like carbon and TiAlSiCN nanocomposite coatings. *Appl. Surf. Sci.* **2011**, *258*, 1206–1211. [CrossRef]
15. Chen, Y.-I.; Lin, Y.-T.; Chang, L.-C.; Lee, J.-W. Preparation and annealing study of CrTaN coatings on WC-Co. *Surf. Coat. Technol.* **2011**, *206*, 1640–1647. [CrossRef]
16. Chen, Y.-I.; Lin, K.-Y.; Wang, H.-H.; Cheng, Y.-R. Characterization of Ta-Si-N coatings prepared using direct current magnetron co-sputtering. *Appl. Surf. Sci.* **2014**, *305*, 805–816. [CrossRef]
17. Chang, L.-C.; Chang, C.-Y.; Chen, Y.-I. Mechanical properties and oxidation resistance of reactively sputtered Ta1-xZrxNy thin films. *Surf. Coat. Technol.* **2015**, *280*, 27–36. [CrossRef]
18. Chen, Y.-I.; Wang, H.-H. Oxidation resistance and mechanical properties of Cr-Ta-Si-N coatings in glass molding processes. *Surf. Coat. Technol.* **2014**, *260*, 118–125. [CrossRef]
19. Lin, T.-N.; Han, S.; Weng, K.-W.; Lee, C.-T. Investigation on the structural and mechanical properties of anti-sticking sputtered tungsten chromium nitride films. *Thin Solid Films* **2013**, *529*, 333–337. [CrossRef]
20. Chen, Y.-I.; Cheng, Y.-R.; Chang, L.-C.; Lee, J.-W. Chemical inertness of Cr-W-N coatings in glass molding. *Thin Solid Films* **2015**, *593*, 102–109. [CrossRef]
21. Zhang, X.; Zhou, Y.-W.; Gao, J.-B.; Zhao, Z.-W.; Guo, Y.-Y.; Xie, Z.-W.; Kelly, P. Effect of the filament discharge current on the microstructure and performance of plasma-enhanced magnetron sputtered TiN coatings. *J. Alloys Compd.* **2017**, *725*, 877–883. [CrossRef]
22. Gu, B.; Tu, J.P.; Zheng, X.H.; Yang, Y.Z.; Penga, S.M. Comparison in mechanical and tribological properties of Cr-W-N and Cr-Mo-N multilayer films deposited by DC reactive magnetron sputtering. *Surf. Coat. Technol.* **2008**, *202*, 2189–2193. [CrossRef]
23. Lin, C.-H.; Duh, J.-G.; Yau, B.-S. Processing of chromium tungsten nitride hard coatings for glass molding. *Surf. Coat. Technol.* **2006**, *201*, 1316–1322. [CrossRef]
24. Hones, P.; Sanjinés, R.; Lévy, F. Sputter deposited chromium nitride based ternary compounds for hard coatings. *Thin Solid Films* **1998**, *332*, 240–246. [CrossRef]

25. Wu, Z.; Tian, X.; Gong, C.; Yang, S.; Chu, P.K. Micrograph and structure of CrN films prepared by plasma immersion ion implantation and deposition using HPPMS plasma source. *Surf. Coat. Technol.* **2013**, *229*, 210–216. [CrossRef]
26. Lippitz, A.; Hübert, T.H. XPS investigations of chromium nitride thin films. *Surf. Coat. Technol.* **2005**, *200*, 250–253. [CrossRef]
27. Chang, J.H.; Jung, M.N.; Park, J.S.; Park, S.H.; Im, I.H.; Lee, H.J.; Ha, J.S.; Fujii, K.; Hanada, T.; Yao, T.; et al. X-ray photoelectron spectroscopy study on the CrN surface grown on sapphire substrate to control the polarity of ZnO by plasma-assisted molecular beam epitaxy. *Appl. Surf. Sci.* **2009**, *255*, 8582–8586. [CrossRef]
28. Zhang, X.X.; Wu, Y.Z.; Mu, B.; Qiao, L.; Li, W.X.; Li, J.J.; Wang, P. Thermal stability of tungsten sub-nitride thin film prepared by reactive magnetron sputtering. *J. Nucl. Mater.* **2017**, *485*, 1–7. [CrossRef]
29. Baker, C.C.; Shah, S.I. Reactive sputter deposition of tungsten nitride thin films. *J. Vac. Sci. Technol. A* **2002**, *20*, 1699–1703. [CrossRef]
30. Wang, Z.; Liu, Z.; Yang, Z.; Shingubara, S. Characterization of sputtered tungsten nitride film and its application to Cu electroless plating. *Microelectr. Eng.* **2008**, *85*, 395–400. [CrossRef]
31. Agouram, S.; Bodart, F.; Terwagne, G. LEEIXS and XPS studies of reactive unbalanced magnetron sputtered chromium oxynitride thin films with air. *J. Electron Spectrosc. Relat. Phenom.* **2004**, *134*, 173–181. [CrossRef]
32. Song, H.Y.; Jiang, H.F.; Liu, X.Q.; Jiang, Y.Z.; Meng, G.Y. Preparation of WO_x-TiO_2 and the Photocatalytic Activity under Visible Irradiation. *Key. Eng. Mater.* **2007**, *336–338*, 1979–1982. [CrossRef]
33. Lu, F.H.; Chen, H.Y.; Hung, C.H. Degradation of CrN films at high temperature under controlled atmosphere. *JVSTA* **2003**, *21*, 671. [CrossRef]
34. Asgary, S.; Hantehzadeh, M.R.; Ghoranneviss, M. Temperature dependence of copper diffusion in different thickness amorphous tungsten/tungsten nitride layer. *Phys. Met. Metall.* **2017**, *118*, 1127–1135. [CrossRef]
35. Chang, Y.Y.; Hsiao, C.Y. High temperature oxidation resistance of multicomponent Cr-Ti-Al-Si-N coatings. *Surf. Coat. Technol.* **2009**, *204*, 992–996. [CrossRef]
36. Zou, H.K.; Chen, L.; Chang, K.K.; Pei, F.; Du, Y. Enhanced hardness and age-hardening of TiAlN coatings through Ru-addition. *Scr. Mater.* **2019**, *162*, 382–386. [CrossRef]
37. Ha, C.; Xu, Y.X.; Chen, L.; Pei, F.; Du, Y. Mechanical properties, thermal stability and oxidation resistance of Ta-doped CrAlN coatings. *Surf. Coat. Technol.* **2019**, *368*, 25–32.
38. Chen, L.; Du, Y.; Mayrhofer, P.H.; Wang, S.Q.; Li, J. The influence of age-hardening on turning and milling performance of Ti-Al-N coated inserts. *Surf. Coat. Technol.* **2008**, *202*, 5158–5161. [CrossRef]
39. Mayrhofer, P.H.; Hörling, A.; Karlsson, L.; Sjoelen, J. Self-organized nanostructures in the Ti-Al-N system. *Appl. Phys. Lett.* **2003**, *83*, 2049–2051. [CrossRef]

© 2020 by the authors. Licensee MDPI, Basel, Switzerland. This article is an open access article distributed under the terms and conditions of the Creative Commons Attribution (CC BY) license (http://creativecommons.org/licenses/by/4.0/).

Article

Investigation for Sidewall Roughness Caused Optical Scattering Loss of Silicon-on-Insulator Waveguides with Confocal Laser Scanning Microscopy

Hongpeng Shang, Degui Sun *, Peng Yu, Bin Wang, Ting Yu, Tiancheng Li and Huilin Jiang

Schools of Optoelectronic Engineering, Science and Mechatronic Engineering, Changchun University of Science and Technology, Changchun 130022, China; shanghongpeng@126.com (H.S.); 13620788904@163.com (P.Y.); dawangbin123@163.com (B.W.); yutingguangxue@163.com (T.Y.); 15568553366@163.com (T.L.); HLJIANG@cust.edu.cn (H.J.)
* Correspondence: sundg@cust.edu.cn

Received: 4 February 2020; Accepted: 29 February 2020; Published: 4 March 2020

Abstract: Sidewall roughness-caused optical loss of waveguides is one of the critical limitations to the proliferation of the silicon photonic integrated circuits in fiber-optic communications and optical interconnects in computers, so it is imperative to investigate the distribution characteristics of sidewall roughness and its impact upon the optical losses. In this article, we investigated the distribution properties of waveguide sidewall roughness (SWR) with the analysis for the three-dimensional (3-D) SWR of dielectric waveguides, and, then the accurate SWR measurements for silicon-on-insulator (SOI) waveguide were carried out with confocal laser scanning microscopy (CLSM). Further, we composed a theoretical/experimental combinative model of the SWR-caused optical propagation loss. Consequently, with the systematic simulations for the characteristics of optical propagation loss of SOI waveguides, the two critical points were found: (i) the sidewall roughness-caused optical loss was synchronously dependent on the correlation length and the waveguide width in addition to the SWR and (ii) the theoretical upper limit of the correlation length was the bottleneck to compressing the roughness-induced optical loss. The simulation results for the optical loss characteristics, including the differences between the TE and TM modes, were in accord with the experimental data published in the literature. The above research outcomes are very sustainable to the selection of coatings before/after the SOI waveguide fabrication.

Keywords: sidewall roughness; optical scattering loss; silicon-on-insulator waveguide

1. Introduction

Higher and higher integration density is the long-standing goal of the optical integrated circuits in the modern optical communications systems [1,2]. Since the beginning of the 21st century, the silicon-on-insulator (SOI) waveguide-based photonic integrated circuits (PIC) have shown unprecedented potential. Therefore, as the main source of optical propagation loss (OPL) of waveguides, the surface roughness-caused optical scattering loss has been attracting much research, and several theoretical models for defining the relations between the optical propagation losses and the sidewall roughness (SWR) of a waveguide are proposed. Illustratively, the average optical loss of SOI strip waveguides is around 0.24 dB/mm [3–5], while that of the small area square SOI waveguides is around 1.3 dB/mm [6]. Thus, it is of paramount importance to be able to quantitatively predict the optical loss on the magnitude and distribution property of waveguide SWR so that we can realize the expected performance specifications of PIC devices.

In the study of the mechanisms of the optical propagation loss caused by waveguide SWR, Marcuse proposed the earliest theory in 1969 in which the optical power ratio at the nonuniform boundary was considered [7]. In 1994, Payne and Lacey proposed a theoretical model for defining the relation between the optical scattering loss and sidewall roughness of a waveguide based on a combination of the spectral density of the nonuniform edge of the waveguide, and the autocorrelation of roughness caused wave scattering, so it was then commonly accepted and referred to as the Payne-Lacey (PL) model [8]. However, both the Marcuse model and PL model have a similar shortcoming that the light wave at the waveguide boundary is not specified, and the waveguide channel is simplified to be a two-dimensional (2-D) structure, where the SWR is assumed to have a uniform distribution. Since the beginning of the 21st century, there has been a strong attempt to extend the PL model into the three-dimensional (3-D) structures and consider the other elements causing the optical loss of waveguide in addition to the SWR [9–12]. In 2005, Barwicz and Haus carried out their theoretical work based on the 3-D interaction between the polarized Poynting vector and the vertical shape of the field (VSF) so that they could give rise to the more detailed simulations of the OPL values of both the low and high index-contrast waveguides [9]. In 2006, Poulton et al. gave a more powerful explanation as that the OPL caused by the SWR was from the two conversions: the conversion from a guided-mode to a radiation mode and the conversion from the radiation mode to a leaky mode, so that they could accurately compute the electric fields with the finite-difference time-domain (FDTD) method [10]. Especially, for the SOI waveguides, in 2008, Schmid et al. employed the non-uniform waveguide boundaries to study the scattering loss [11], and, in 2009, Yap et al. first led the optical scattering loss of SOI waveguides to the synchronous dependences the SWR and the channel width [12]. Until 2019, we first published the measurement metrology of the waveguide SWR with a confocal laser scanning microscopy (CLSM) technique [13].

In this article, we theoretically investigated the SWR property of dielectric waveguides by analyzing its components in the horizontal and vertical direction. Then, we measured the 3-D distribution of waveguide SWR with the CLSM technique and constructed a theoretical/experimental ensemble model for defining the SWR-caused optical propagation loss. This model was an extended version of the PL model with the 3-D distribution of SWR. As a result, with such a combinative model, we simulated the dependences of the optical propagation loss coefficient of the waveguide on the SWR distribution and the width. Finally, we analyzed the theoretical and experimental results.

2. Theoretical Analysis for Waveguide Sidewall Roughness

2.1. Analysis for the PL Model

An equivalent rectangular 3-D vision of an SOI waveguide is shown in Figure 1a, where n_1 and n_2 are the refractive indices of the core and cladding layers, respectively, and $2d$ is the width of the waveguide. The SWR of a waveguide was caused by the inductively coupled plasma (ICP)-etching, as shown in Figure 1b, then as shown in Figure 1c, the distributions of the SWR at the vertical and horizontal directions were probably different from each other.

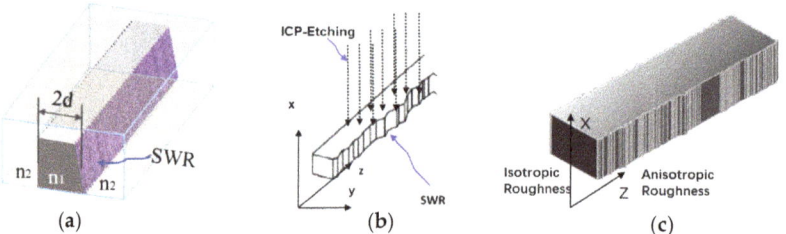

Figure 1. Sidewall roughness (SWR) of silicon-on-insulator (SOI) waveguide: (**a**) the schematic three-dimensional (3-D) equivalent rectangular waveguide with SWR; (**b**) the ICP-etching to a waveguide sidewall; (**c**) the distributions of the SWR at x- and z-coordinate.

In accordance with the 3-D simulations of radiation mode, the three linear polarization states (x, y, and z) could cause the different profiles of the radiation mode [8,9]. The exponential form of the PL model was used to define a space distribution Equation (1) for an interest. The power spectral density of the roughness was of interest in the optical scattering loss, then the Fourier transform of the autocorrelation function of the roughness was expressed to be Equation (2) [8,9]:

$$R(u) \approx \sigma^2 \exp(-|u|/L_c)) \tag{1}$$

$$R(\xi) \approx 2\sigma^2 L_c / (1 + L_c^2 \xi^2) \tag{2}$$

where σ is the root-mean-square (rms) roughness, L_c is the correlation length of the roughness with the assumption that there is no correlation between the two sidewalls, u is the space variable of a waveguide sidewall, and ξ is the space-frequency of the power spectral function of roughness. If the wavelength of the light wave in air was λ, with the wavenumber as $k_0 = 2\pi/\lambda$ and the finite difference processing of beam propagation method (FD-BPM), we obtained the effective index N_{eff} of a single guided-mode, and further with the propagation constant of this guided-mode $\beta = k_0 \cdot N_{eff}$, we cited three dimensionless parameters h, V, and p of guided-mode defined by the PL model as [9]

$$h = d\sqrt{n_1^2 k_0^2 - \beta^2}, \quad V = k_0 d \sqrt{n_1^2 - n_2^2} \text{ and } p = d\sqrt{\beta^2 - n_2^2 k_0^2} \tag{3}$$

where h and p are the very popular parameters, defining the guided mode field in the literature on optical waveguides [1,2]. V is the product of three elements: the numerical aperture $\sqrt{n_1^2 - n_2^2}$ of a symmetric planar waveguide, d is the radius (or half a width) of the waveguide core, and k_0 is the wave number in the air. Further, we obtained the dimensionless parameters as:

$$\Delta = (n_1^2 - n_2^2)/(2n_1^2), \quad x = p(L_c/d), \quad \gamma = (n_2 V)/(n_1 p \sqrt{\Delta}) \tag{4}$$

Consequently, for the two-SWR-induced optical scattering loss, with the above definitions for the guided-mode profile defined by Equations (3) and (4) and a combination of the PL model (1)–(4), the Yap improvement for the optical loss coefficient dependence on the SWR was expressed as [12]

$$\alpha_{PL}(TE/TM) = \frac{4.34 \sigma_{2D}^2}{\sqrt{2} d^4 \beta_{TE/TM}} g(V) \cdot f_e(x, \gamma) \tag{5}$$

where σ_{2D} is the SWR defined in the 2-D form, the loss coefficient is in dB/cm, and the functions $g(V)$ and $f_e(x,\gamma)$ are defined by

$$g(V) = \frac{h^2 V^2}{1+p^2} \text{ and } f_e(x,\gamma) = \frac{\{[(1+x^2)^2 + 2x^2\gamma^2]^{1/2} + 1 - x^2\}^{1/2}}{[(1+x^2)^2 + 2x^2\gamma^2]^{1/2}} \tag{6}$$

The model defined by Equations (5) and (6) might be referred to as a Yap-form PL model.

2.2. Three-Dimensional Model for the SWR-caused Optical Scattering Loss

In the roughness-improved PL model, the correlation length of sidewall roughness L_c is a paramount important parameter [9]. Before 2000, for a low index-contrast waveguide, such as silica-waveguide, L_c was observed to be a few micrometers, and even the values less than 500 nm were ever exploited, and meanwhile, the optical scattering loss was determined to be a forward-scattering process, where $L_c \approx 1/2\beta$ that matched with the maximum attenuation [14]. In contrast, Barwicz and Haus studied the 3-D optical scattering process for both the high and low index-contrast waveguides with respect to three L_c values as 1, 50, and 150 nm [9]. In the Barwicz-Haus's 3-D theory, the optical scattering loss is thought to be caused by a radiative mode coupled from a guided mode, meanwhile, for straight roughed waveguides, the phase-matching condition between the guided and the radiated modes allows only a narrow range of spectral frequencies of roughness to produce radiation loss, then the estimated correlation length L_c value is in a range of $1/(\beta + n_2 k_0) < L_c < 1/(\beta - n_2 k_0)$. Consequently, for both the TE-like and TM-like modes, the SWR-induced optical scattering loss was related to both the y- and z-components of roughness, so the dependences of the optical intensity loss coefficients $\alpha_{3D}(TE/TM)$ on the SWR $\sigma_{3D}(TE/TM)$ could be expressed as

$$\alpha_{3D}(TE/TM) = 4.34 \left(\frac{\sigma_{3D}^2(TE/TM)}{\sqrt{2}d^4 \beta_{TE/TM}} g(V) \cdot f_e(x, \gamma) \right) \quad (7)$$

Hence, the new model Equation (7) was the combination of the Yap-form PL model and the 3-D SWR distribution. In this model, the optical loss coefficient was in dB/cm.

3. Measurements for the Roughness Characteristics with CLSM

3.1. Experimental Measurement

Figure 2 shows the surface roughness measurement mechanism with a CLSM system: (a) is a schematic optical imaging system in which the laser beam as a probe has a wavelength of 405 nm, and the section of scanning is set 10 nm, and the spot size of the focused laser beam is defined by the full width at the half-maximum (FWHM) of the laser intensity distribution, (b) is an illustrative sample of the retrieved image in the CLSM measurement, and (c) is the standard definition of the peak-to-valley (P-V) roughness.

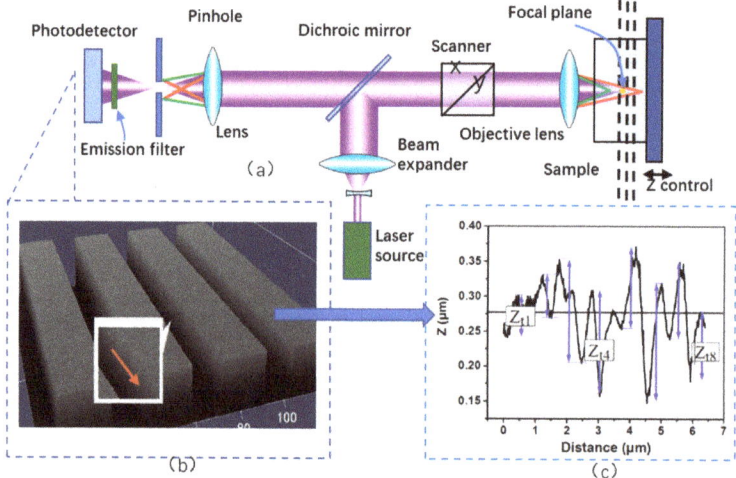

Figure 2. CLSM (confocal laser scanning microscopy) measurement and the definition for surface roughness: (**a**) The schematic optical imaging system, (**b**) the retrieved image of CLSM-measured result of an object, and (**c**) the definition for the fabricated surface roughness.

In the CLSM metrology, the roughness is defined as a root mean square (rms) PS_z of all the P-V periods as [13,15]

$$PS_z = (1/N) \sqrt{\left(\sum_{i=1}^{N} R_{ai}^2 \right)} \quad (8)$$

where R_{ai} is the average roughness at the ith scanned period in the CLSM measurement. In this definition of roughness, the PS_z could be taken as a one-dimensional (1-D) roughness σ_{1D}, then the 2-D average roughness σ_{2D} and the 3-D average roughness σ_{3D} are defined as

$$\sigma_{2D} = (1/m) \sum_{j=1}^{m} PS_z(aj) \tag{9a}$$

$$\sigma_{3D} = (1/n) \sum_{k=1}^{n} \sigma_{2D}(ak) \tag{9b}$$

3.2. Experimental Measurements for the SWR of SOI Waveguide Sample

In this work, we selected the commercial CLSM tool–LSM710 that was produced by the ZEISS company. Then, the device sample for the CLSM measurements was selected by selecting an SOI-waveguide sample having a BOX layer of 2.0 and 1.5 µm, where the waveguide ribs having a width of 4.0 µm and a height of 0.5 µm were etched with the advanced ion etching—the inductively coupled plasma (ICP) etching technique. The CLSM measurements only showed the existing isotropic roughness, then we obtained the reconstructed image of the waveguide SWR, as shown in Figure 3a, and by scanning the measured area along a vertical direction at the $x = 100$ nm position, a line-scanning roughness was obtained as $P_z = 20$ nm from the data display. Then, in the same manner, we measured the other five lines with every 100 nm position change along the vertical direction for two reconstructed images (the total height of the etched SOI rib was 500 nm), and then the total 6 measured values are depicted in Figure 3b, which were in the range of 16–23 nm roughness and gave rise to an average isotropic SWR of 20.33 nm. First, we needed to clarify that the x coordinate in Figure 3 was the traveling direction z of the optical beam, the z and y coordinates of Figure 3 were the coordinates x and z of Figure 1, respectively. Then, we also noticed that the SWR values of the SOI waveguides were only in the range of a few tens of nanometers, which were much smaller than the CLSM-measured values of the silica waveguide etched by the traditional reactive ion etching (RIE) technique owing to the advanced etching technique ICP and the etched material of silicon [16,17]. Accordingly, the disparities of SWR distributions between the x and y coordinates existed.

Here, what needed to be clarified was that the measurement accuracy was 10 nm in the above measurements, but for the roughness smaller than 10 nm, the accuracy needed to be improved, which would be realized with the improvement of the laser FWHM values in both the lateral and axial directions. In addition, as shown in Figure 2a, in one CLSM measurement, we scanned the device sample having four equal size waveguides to acquire the data, but as shown in Figure 3a, only one waveguide was selected from the reconstructed CLSM image to analyze and measure the roughness. In order to analyze the SWR uniformity of one channel, several sections could be selected along the waveguide to obtain the average SWR value of each section, and then all the roughness results of all the sections could be compared. In the same manner, in order to analyze the SWR uniformity among all the four measured waveguides, the CLSM measured results of SWR for all the four waveguides could be carried out, and then all the roughness results could be compared.

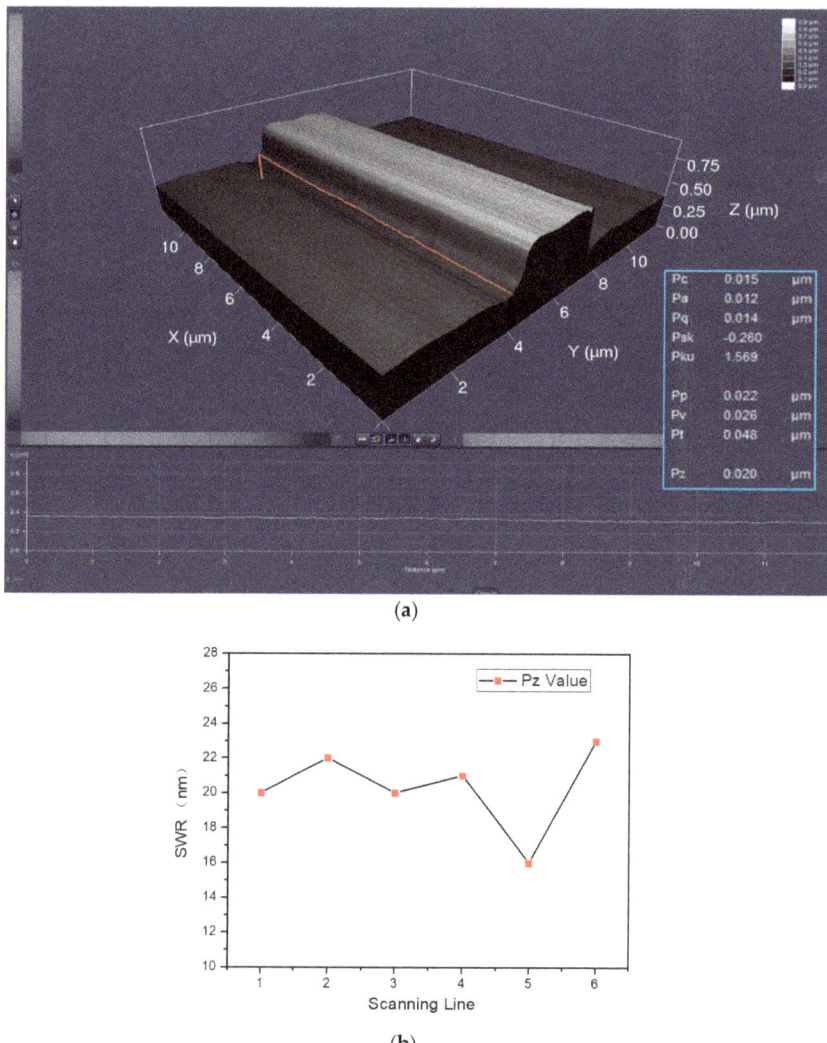

(a)

(b)

Figure 3. CLSM Measurements and data process of SOI waveguide SWR: (**a**) is the reconstructed image of CLSM measurement, (**b**) is the isotropic roughness distribution of five measured spots.

4. Verifications for the SWR-caused Optical Propagation Loss

4.1. Simulations for the SWR-Caused Optical Scattering Loss

In the CLSM system shown in Figure 2, the measurements shown in Figure 3 presented that an SOI waveguide generally had much lower SWR value than silicon dioxide (SiO_2) waveguides, even when they were fabricated under the same etching technique, ICP. Thus, it could be forecasted that the waveguide dimensions of both core and cladding layers of a waveguide system probably have significant effects on the optical scattering loss apart from the sidewall roughness itself.

Based on the optical performance of the real SOI waveguide functional device, we selected an SOI waveguide sample having the rib width and height of 4.0 and 0.5 μm, respectively; at 1550 nm wavelength, the refractive index silicon film was 3.4777 [18], and the refractive index of both the

BOX layer and upper cladding layer was 1.4394 and 1.4449 for TE and TM modes, respectively; then, the effective indices of 3.3254 and 3.3168, respectively, were obtained with the simulation of beam propagation method (BPM) software. Further, by selecting the 2-D SWR construction defined by Equation (9a) and with the improved model Equation (7), we obtained the simulation results of the SWR dependences of the OPL coefficient, as shown in Figure 4a. Note from Figure 4a that the SWR-induced OPL of TM-mode was higher than that of TE mode for the SOI waveguide, which was not consistent with the results published in the literature [11,12]. Thus, it turned out that both the core and cladding layers of the waveguide system probably had significant effects on the optical scattering loss apart from the sidewall roughness itself. In contrast, by selecting the 3-D SWR construction defined by Equation (9b) and with the improved model Equation (7) of the SWR-caused OPL, we obtained the simulation results of the roughness dependences of the OPL coefficient for the same waveguide sample, as shown in Figure 4b. Note from Figure 4b that with the consideration of the 3-D construction of the roughness, there were two different points of the OPL coefficient between TE and TM modes. One was that the absolute TE-TM difference of the 3-D roughness was relatively larger than that of the 2-D roughness, and the other was that the OPL of TE mode was higher than that of TM mode, which was inverse to the 2-D roughness, but it was really in accord with the measured values published in the literature [11,12]. However, for the measured SWR value in Figure 3, ~20 nm, the OPL values for both the 2-D and 3-D SWR constructions were in the range of 3–3.5 dB/cm.

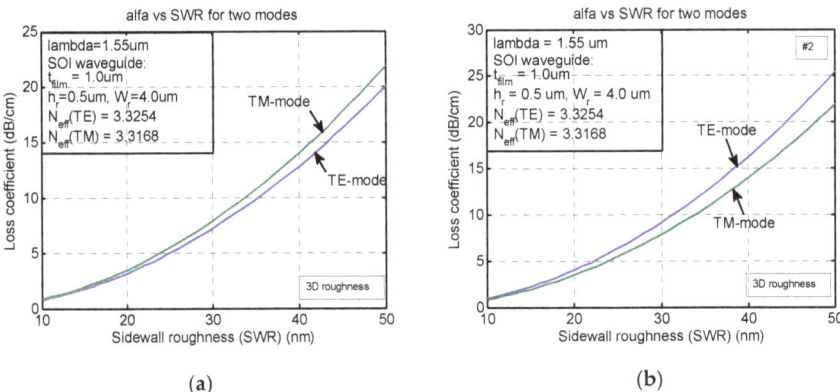

Figure 4. Simulations for the optical loss coefficient of SOI waveguide vs. SWR for two SWR constructions: (**a**) the 2D SWR, (**b**) the 3D-SWR.

As mentioned above, the measurements showed that the SOI-waveguides sample fabricated in the ICP technique only had an average roughness of 20 nm, namely, the 2-D statistic values at both the horizontal and vertical directions were the same. Then, with the 3-D roughness construction and the same values of the SOI waveguide parameters as used for the simulations in Figure 4b, we simulated the synchronous dependences of the roughness-induced optical propagation loss coefficient on both the correlation length L_c and the rib width $W_r(2d)$ of the waveguide, as shown in Figure 5.

The most impressive finding from Figure 5 was that once the SWR values were given, the correlation length L_c had the most dominant impact, and the waveguide width $2d$ had the nonignorable impact upon the roughness-induced OPL in addition to the SWR itself. Accordingly, based on the optimal states of both the SWR and geometrical configuration of a waveguide, to significantly increase the correlation length L_c of the waveguide was the most effective metrology to completely solving the OPL of SOI waveguides. However, the L_c is a function pertaining to the SWR and the dimension of the waveguide [8].

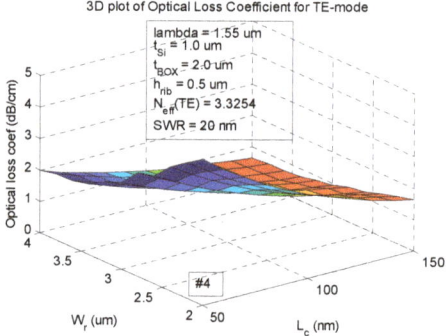

Figure 5. Simulations for the dependences of optical loss of SOI waveguide on both the correlation length and the waveguide width apart from SWR itself of the TE-mode of the waveguide.

4.2. Experimental Measurements for the Optical Propagation Loss of an SOI Waveguide

Among the methods for measuring the OPL of waveguides, the Fabry–Perot (F-P) resonance method is the most appropriate approach when the SOI waveguide chip has a length of around 1 cm due to the ultrahigh fiber-waveguide but coupling loss [18]. With this method, the two ends of the waveguide were polished to form an optical F-P cavity for an optical signal launched into the waveguide. When the optical length of the cavity was continuously changed with either the refractive index of the waveguide material or the wavelength of the optical signal, the optical intensity transmission coefficient of resonance output of the cavity would present a form of the periodic wave. In this method, if the wavelength λ, the waveguide length l_{WG}, and the intensity reflection coefficient R_{end} of two waveguide ends were given, the resonance output intensity of the waveguide cavity was defined by Feuchter and Thirstrup in 1994 [19]. Then, with the special transmittance values, T_{max} and T_{min}, its optical propagation loss α coefficient was obtained. In the experiments, we uniformly changed the refractive index of an SOI waveguide by heating the waveguide with a length of 8.5 mm, then the optical resonant output curve with the temperature was obtained, as shown in Figure 6, and, consequently, the optical propagation loss of $\alpha = \sim 3.1$ dB/cm was reached. In the SWR measurement and prediction of the SWR-caused OPL, this result was very agreeable with the numerical simulation results, shown in Figures 4 and 5, in which the average 20.33 nm anisotropic SWR based on the CLSM measurements were exploited and the OPL distributions of 3.0–3.5 dB/cm that were obtained with the 3-D construction in Figure 4b. From the view of the fabrication of SOI waveguides, the OPL of 3.1 dB/cm was relatively lower than the value of 3.6 ± 0.1 dB/cm, published in 2002 [20], and relatively higher than the value of around 1.4 dB/cm, published in 2016 [21]. So, the ICP fabrication of the waveguide samples employed in this work still needs to be improved.

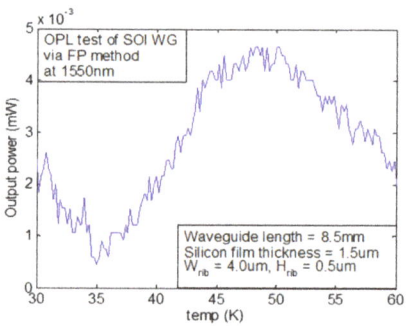

Figure 6. Measurement output curve of Fabry–Perot (F-P) resonance of the SOI waveguide vs. temperature.

5. Conclusions

In this work, the analyses for the ion etching-based SWR of a rectangular dielectric waveguide led to the categorization of the horizontal and vertical components, and the P-V definition of type surface roughness of CLSM measurements also showed the ability to scan the etched surface at the two directions. Then, the possible different contributions of two components to OPL of waveguide could be imaged and reconstructed as the measurement results. Therefore, with the accurate measured values of the SWR and the theoretical/experimental combinative model, the accurate OPL coefficient was composed, and the important simulation results of the SWR-caused optical loss were found with this model. Therefore, the conclusions obtained in this article are very sustainable in considering the ensemble effect of SWR and the SOI waveguide structure in the research and development of high-quality industrial devices and systems. As an additional conclusion from this work, the upper limit of the correlation length was calculated to be 130 nm, and in any similar work in the future, this parameter is necessary to be discussed. With the above simulation results and conclusions, the coating works of silicon dioxide films of both the BOX layer and the upper cladding layer could be specifically designed to compress the optical loss.

Author Contributions: Conceptualization, H.S. and D.S.; methodology, H.S., P.Y., B.W. and T.L.; validation, H.S. and D.S.; formal analysis, D.S., H.S. and H.J.; investigation, D.S., H.S. and T.Y.; data curation, H.S. and B.W; writing—original draft preparation, H.S.; writing—review and editing, H.J. and D.S.; visualization, P.Y. All authors have read and agreed to the published version of the manuscript.

Funding: This work is co-sponsored by the Natural Science Foundation of Jilin Provincial Science & Technology (Foundation Grant: 20180101223JC).

Acknowledgments: The authors thank the Chang Guang Yuanchen Optoelectronics, Ltd., for their help in Si etching and PECVD coatings of SOI wafer samples. Else, the authors would like to thank Guozheng Wang of the School of Science at the CUST, China, for his help in wafer processing, and want to thank a graduate, Xueping Wang, for her support in testing.

Conflicts of Interest: The authors declare no conflict of interest.

References

1. Doerr, C.R.; Okamoto, K. Planar lightwave circuits in fiber-optic communications. In *Optical Fiber Telecommunications*; Kaminow, I., Li, T., Eds.; Academic Press/Elsevier: New York, NY, USA, 2008.
2. Murphy, E.J. *Optical Circuits and Components: Design and Applications*; Marcel Dekker, Inc.: New York, NY, USA, 1999.
3. Chrostowski, K.; Hochberg, M. Silicon photonics design: From devices to systems. In *Silicon Photonics: Fueling the Next Information Revolution*, 1st ed.; Innis, D., Rubenstein, R., Eds.; Cambridge University Press: Cambridge, UK, 2015; pp. 3–27.
4. Orcutt, J.S.; Moss, B.; Sun, C.; Leu, J.; Georgas, M.; Zgraggen, J.S.E.; Li, H.; Sun, J.; Weaver, M.; Uroševic, S.; et al. Open foundry platform for high-performance electronic-photonic integration. *Opt. Express* **2012**, *20*, 12222–12232. [CrossRef] [PubMed]
5. Dumon, P.; Bogaerts, W.; Wiaux, V.; Wouters, J.; Beckx, S.; Van Campenhout, J.; Taillaert, D.; Luyssaert, B.; Bienstman, P.; Van Thourhout, D.; et al. Low-loss SOI photonic wires and ring resonators fabricated with deep UV lithography. *IEEE Photonics Technol. Lett.* **2004**, *16*, 1328–1330. [CrossRef]
6. Tsuchizawa, T.; Yamada, K.; Fukuda, H.; Watanabe, T.; Takahashi, J.I.; Takahashi, M.; Shoji, T.; Tamechika, E.; Itabashi, S.I.; Morita, H. Microphotonics devices based on silicon microfabrication technology. *IEEE J. Sel. Top. Quantum Electron.* **2005**, *11*, 232–240. [CrossRef]
7. Marcuse, D. Radiation losses of dielectric waveguides in terms of the power spectrum of the wall distortion function. *Bell Syst. Technol. J.* **1969**, *48*, 3233–3244. [CrossRef]
8. Payne, F.P.; Lacey, J.P.R. A theoretical analysis of scattering loss from planar optical waveguides. *Opt. Quant. Electron.* **1994**, *26*, 977–986. [CrossRef]
9. Barwicz, T.; Haus, H.A. Three-dimensional analysis of scattering loss due to sidewall roughness in micro photonic waveguides. *IEEE J. Lightw. Technol.* **2005**, *3*, 2719–2732. [CrossRef]

10. Poulton, C.G.; Koos, C.; Fujii, M.; Pfrang, A.; Schimmel, T.; Leuthold, J.; Freude, W. Radiation models and roughness loss in high index-contrast waveguides. *IEEE J. Sel. Top. Quantum Electron.* **2006**, *12*, 1306–1321. [CrossRef]
11. Schmid, J.H.; Delâge, A.; Lamontagne, B.; Lapointe, J.; Janz, S.; Cheben, P.; Densmore, A.; Waldron, P.; Xu, D.-X.; Yap, K.P. Interference effect in scattering loss of high-index-contrast planar waveguides caused by boundary reflections. *Opt. Lett.* **2008**, *33*, 1479–1481. [CrossRef] [PubMed]
12. Yap, K.P.; Delâge, A.; Lapointe, J.; Lamontagne, B.; Schmid, J.H.; Waldron, P.; Syrett, B.A.; Janz, S. Correlation of scattering loss, sidewall roughness and waveguide width in silicon-on-insulator (SOI) ridge waveguides. *IEEE J. Lightw. Technol.* **2009**, *27*, 3999–4007.
13. Sun, D.G.; Shang, H.; Jiang, H.L. Effective metrology and standard of the surface roughness of micro/nanoscale waveguides with confocal laser scanning microscopy. *Opt. Lett.* **2019**, *44*, 747–750. [CrossRef] [PubMed]
14. Lee, K.K.; Lim, D.R.; Luan, H.-C.; Agarwal, A.; Foresi, J.; Kimerling, L.C. Effect of size and roughness on light transmission in a Si/SiO$_2$ waveguide: Experiments and model. *Appl. Phys. Lett.* **2000**, *39*, 2705–2718. [CrossRef]
15. Shang, H.; Sun, D.G.; Sun, Q.; Yu, P.; Gao, J.; Hall, T.J. Analysis for system errors in measuring sidewall angle of silica waveguides with confocal laser scanning microscope (CLSM). *Meas. Sci. Technol.* **2019**, *30*, 025004. [CrossRef]
16. Cardinaud, C.; Peignon, M.C.; Tessier, P.Y. Plasma etching: Principle, applications to micro- and nano-technologies. *Appl. Surf. Sci.* **2000**, *164*, 72–83. [CrossRef]
17. Brault, P.; Dumas, P.; Salvan, F. Roughness scaling of plasma-etched silicon surface. *J. Phys. Cond. Matter* **1998**, *10*, 27–38. [CrossRef]
18. Dwivedi, S.; Ruocco, A.; Vanslembrouck, M.; Spuesens, T.; Bienstman, P.; Dumon, P.; Van Vaerenbergh, T.; Bogaerts, W. Experimental extract of effective refractive index and thermo-optic coefficients of silicon-on-insulator waveguides using interferometer. *IEEE J. Lightwave Technol.* **2015**, *33*, 4471–4477. [CrossRef]
19. Feuchter, T.; Thirstrup, C. High precision planar waveguide propagation loss measurement technique using a Fabry-Perot cavity. *IEEE Photonics Technol. Lett.* **1994**, *6*, 1244–1247. [CrossRef]
20. Vlasov, V.A.; McNab, S.J. Losses in single-mode silicon-on-insulator strip waveguides and bends. *Opt. Express* **2002**, *12*, 1622–1631. [CrossRef] [PubMed]
21. Boeuf, F.; Cremer, S.; Temporiti, E.; Fere, M.; Shaw, M.; Baudot, C.; Vulliet, N.; Pinguet, T.; Mekis, A.; Masini, G.; et al. Silicon photonic R&D and manufacturing on 300-mm wafer platform. *IEEE J. Lightwave Technol.* **2016**, *33*, 286–295.

© 2020 by the authors. Licensee MDPI, Basel, Switzerland. This article is an open access article distributed under the terms and conditions of the Creative Commons Attribution (CC BY) license (http://creativecommons.org/licenses/by/4.0/).

MDPI
St. Alban-Anlage 66
4052 Basel
Switzerland
Tel. +41 61 683 77 34
Fax +41 61 302 89 18
www.mdpi.com

Coatings Editorial Office
E-mail: coatings@mdpi.com
www.mdpi.com/journal/coatings

www.ingramcontent.com/pod-product-compliance
Lightning Source LLC
LaVergne TN
LVHW070554100526
838202LV00012B/463